IEE CIRCUITS AND SYSTEMS SERIES 4

Series Editors: Dr D. G. Haigh
Dr R. Soin

ALGORITHMIC
and
KNOWLEDGE BASED
CAD
for
VLSI

Other volumes in this series:

Volume 1 **GaAs technology and its impact on circuits and systems**
D. G. Haigh and J. Everard (Editors)
Volume 2 **Analogue IC design: the current-mode approach**
C. Toumazou, F. J. Lidgey and D. G. Haigh (Editors)
Volume 3 **Analogue-digital ASICs—circuit techniques, design tools and applications**
R. S. Soin, F. Maloberti and J. Franca (Editors)

ALGORITHMIC and KNOWLEDGE BASED CAD for VLSI

Edited by
GAYNOR TAYLOR
and
GORDON RUSSELL

Peter Peregrinus Ltd. on behalf of the Institution of Electrical Engineers

Published by: Peter Peregrinus Ltd., London, United Kingdom

© 1992: Peter Peregrinus Ltd.

Apart from any fair dealing for the purposes of research or private study, or criticism or review, as permitted under the Copyright, Designs and Patents Act, 1988, this publication may be reproduced, stored or transmitted, in any forms or by any means, only with the prior permission in writing of the publishers, or in the case of reprographic reproduction in accordance with the terms of licences issued by the Copyright Licensing Agency. Inquiries concerning reproduction outside those terms should be sent to the publishers at the undermentioned address:

Peter Peregrinus Ltd.,
Michael Faraday House,
Six Hills Way, Stevenage,
Herts. SG1 2AY, United Kingdom

While the editors and the publishers believe that the information and guidance given in this work is correct, all parties must rely upon their own skill and judgment when making use of it. Neither the editors nor the publishers assume any liability to anyone for any loss or damage caused by any error or omission in the work, whether such error or omission is the result of negligence or any other cause. Any and all such liability is disclaimed.

The right of the editors to be identified as editors of this work has been asserted by them in accordance with the Copyright, Designs and Patents Act 1988.

British Library Cataloguing in Publication Data

A CIP catalogue record for this book
is available from the British Library

ISBN 0 86341 267 X

Printed in England by Short Run Press Ltd., Exeter

Contents

Foreword	ix
Preface	x
List of contributors	xii

1 Expert assistance in digital circuit design — 1
  1.1 Introduction — 1
  1.2 Digital systems — 1
  1.3 Present CAD systems — 2
  1.4 The nature of design — 4
  1.5 Design is a 'satisfactory' process — 6
  1.6 The artificial intelligence paradigm — 7
  1.7 The design assistant — 10
    1.7.1 General scheme of a design automation system — 11
    1.7.2 Aims of the design assistant — 14
    1.7.3 Requirements for the design assistant — 16
    1.7.4 Present status — 18
    1.7.5 Functional arrangement of the design assistant — 19
    1.7.6 The knowledge base — 19
  1.8 Conclusion — 21
  1.9 References — 21

2 Use of a theorem prover for transformational synthesis — 24
  2.1 Introduction — 24
  2.2 Transformational synthesis — 25
  2.3 The Boyer-Moore theorem prover — 26
  2.4 A proof example — 29
  2.5 Synthesis method — 32
  2.6 Synthesis example — 37
  2.7 Conclusions — 45
  2.8 References — 45

3 An overview of high level synthesis technologies for digital ASICs — 47
  3.1 Introduction — 47
  3.2 The benefits of synthesis — 50
  3.3 System level synthesis — 51
  3.4 Interface synthesis — 51
  3.5 High level synthesis — 52
    3.5.1 Translation to intermediate representation — 52
    3.5.2 Optimisation — 57
    3.5.3 Using estimation to meet constraints — 61

vi Contents

	3.5.4 Formal methods	62
	3.5.5 Timing constraints and analysis	63
	3.5.6 Algorithmic techniques	63
3.6	Synthesis specification language considerations	67
	3.6.1 General requirements	67
	3.6.2 Using VHDL for synthesis	67
	3.6.3 Other synthesis languages	69
3.7	Summary	70
3.8	Acknowledgments	71

4 Simulated annealing based synthesis of fast discrete cosine transform blocks — 75
- 4.1 Introduction — 75
- 4.2 Problem domain — 75
- 4.3 Synthesis and simulated annealing — 76
  - 4.3.1 The behavioural synthesis procedure — 78
  - 4.3.2 The simulated annealing algorithm — 79
- 4.4 Test results — 86
  - 4.4.1 The structural synthesis tools — 87
- 4.5 Conclusions — 89
  - 4.5.1 Current developments — 89
  - 4.5.2 Concluding remarks — 90
- 4.6 Acknowledgments — 91
- 4.7 References — 91

5 Knowledge based expert systems in testing and design for testability—an overview — 94
- 5.1 Introduction — 94
- 5.2 Components of a knowledge based expert system — 96
- 5.3 Test generation — 101
- 5.4 Design of testability — 107
- 5.5 Conclusions — 118
- 5.6 References — 119

6 Knowledge based test strategy planning — 122
- 6.1 Introduction — 122
- 6.2 Test strategy planning — 122
- 6.3 Knowledge based systems and test planning — 124
  - 6.3.1 Current systems — 124
  - 6.3.2 Test strategy description and evaluation — 126
- 6.4 Cost modelling and test strategy evaluation — 126
  - 6.4.1 The test method knowledge base — 128
  - 6.4.2 Test strategy evaluation — 129
- 6.5 Test method selection — 132
- 6.6 The test strategy controller — 137
- 6.7 Conclusions — 140
- 6.8 References — 140

7 HIT: a hierarchical integrated test methodology — 142
- 7.1 Introduction — 142
- 7.2 System overview — 145
  - 7.2.1 Cell characterisation — 146
  - 7.2.2 Design testability estimators — 149

Contents   vii

		7.2.3 Design partitioning	150
		7.2.4 High level test generation and design modification	151
	7.3	Conclusions	160
	7.4	References	160

8	Use of fault augmented functions for automatic test pattern generation	163
	8.1 Introduction	163
	8.2 The fault augmented function (FAF)	164
	8.3 The original algorithm	165
	8.4 Generating the FAFs	165
	8.5 Problems with FAF generation	168
	8.6 Implications of the single fault assumption	169
	8.7 Extracting tests from FAFs	174
	8.8 Current state of the software	176
	8.9 Appraisal of the FAF technique	177
	8.10 Future work	179
	8.11 References	179

9	Macro-test: a VLSI testable-design technique	180
	9.1 Introduction	180
	9.2 Testability	180
	9.3 Principles of macro-test	183
	9.4 A test plan: the macro access protocol	185
	9.5 DFT rules for macro-testability	187
	9.5.1 Leaf-macro testability rules	187
	9.5.2 DFT rules for test data access	189
	9.5.3 DFT rules for test control access	191
	9.6 The test control block concept	191
	9.6.1 Introduction	191
	9.6.2 TCBs in hierarchical designs	192
	9.7 Sphinx software overviews	193
	9.7.1 The Sphinx tools	193
	9.7.2 The Sphinx user interface	196
	9.8 Sphinx application examples	196
	9.9 Acknowledgment	199
	9.10 References	199

10	An expert systems approach to analogue VLSI layout	201
	10.1 Introduction and background	201
	10.2 Description of the system	203
	10.2.1 Complete solution	208
	10.2.2 Partial solution	208
	10.2.3 No solution found	209
	10.3 Implementation of the general concept	210
	10.3.1 General placement concept	210
	10.3.2 Heuristic decision evaluation for placement	214
	10.3.3 Component orientation	218
	10.3.4 Grouping of components	223
	10.3.5 Routing between interconnects	225
	10.3.6 Layer-type definition	234
	10.4 Comments and conclusions	240
	10.5 References	242

viii Contents

11 Guaranteeing optimality in a gridless router using AI techniques — 245
   11.1 Introduction — 245
   11.2 Routing generation — 246
   11.3 The gridless single-net routing tool — 251
      11.3.1 Model definition — 251
      11.3.2 Representation of the continuous domain — 252
      11.3.3 A method for integrated search and graph generation — 254
      11.3.4 A practical multi-layer manhattan router — 257
   11.4 Optimisation criteria — 260
   11.5 Operation of the routing tool — 263
   11.6 Conclusions — 266
   11.7 Acknowledgment — 267
   11.8 References — 267

Index — 269

Foreword

This is the fourth volume in the Circuits and Systems and arose out of a colloquium on CAD tools organised by the IEE. The editors and authors have carefully enhanced the material both in terms of the depth of treatment and the breadth of material covered by the original colloquium.

This volume has followed a similar trajectory to the previous ones in this series by accompanying a tutorial course run in conjunction with an international conference, viz. ECCTD '91 (European Conference on Circuit Theory and Design), in Copenhagen, Denmark.

The unrelenting increase in size and complexity of VLSI demands a commensurate development of CAD tools aimed at the various phases of the entire design cycle; ranging from synthesis and functional verification to testability analysis, test generation and fault coverage and layout. The majority of commercially available tools are based on an algorithmic approach and much research effort is aimed at improvement of existing tools and the development of new algorithms. The magnitude of the problem has led to an interest in examining the applicability of expert system and other knowledge based techniques to certain problems in this area and a number of results are becoming available.

Therefore the present volume will make a timely appearance in this field and be a valuable contribution to increasing awareness of the state of the art of CAD tools for VLSI.

Randeep Soin
David Haigh
Fareham, November, 1991

Preface

The continuing growth in the size and complexity of VLSI devices requires a parallel development of well-designed, efficient CAD tools. Such tools must be available for the whole design cycle - synthesis and functional verification, testability analysis, test generation and fault coverage and layout. The majority of commercially available tools are based on an algorithmic approach to the problem and there is a continuing research effort aimed at improving existing tools of this form and developing new algorithms. The sheer complexity of the problem has, however, led to an interest in examining the applicability of expert system and other knowledge based techniques to certain problems in the area and a number of results are becoming available. The aim of this book, which is based on material given at an IEE Colloquium of the same name, is to sample the present state-of-the-art in CAD for VLSI and it covers both newly developed algorithms and applications of techniques from the artificial intelligence community. The editors believe it will prove of interest to all engineers concerned with design and test of integrated circuits and systems. Although it is not intended as a course text many of the chapters provide interesting background reading for postgraduate and final year undergraduate students following courses in VLSI design.

Chapters are arranged in three groups covering topics in synthesis, test and testability and layout. The first section on synthesis comprises 4 chapters, the first of these describes an 'expert assistant' which can perform both design and analysis of microprocessor based systems consisting of 'off-the-shelf' components. The use of formal methods in synthesis is a topic of growing research interest, particularly in the design of 'safety-critical' systems. Chapter 2 describes the formal methods approach to performing, efficiently and correctly, the transformations necessary in synthesising a high level description of a system down to layout. Chapter 3 is a comprehensive overview of synthesis techniques starting with the system level and including translation to intermediate representations and consideration of various constraints and ways of including these in the design. The final chapter in this section discusses the application of simulated annealing methods to the synthesis of function blocks in the area of signal processing with the objective of producing more efficient realisations of functions within given cost and time constraints. The section on test and testability opens with Chapter 5 which gives an overview of the use of knowledge based systems in this area; this is followed by a description of a knowledge based test strategy planner with an emphasis on economic considerations. Chapters 7 and 8 describe tools and techniques to allow design for test and test strategies to be incorporated as an integral part of the design process. The last Chapter in this section outlines an algebraic algorithm to generate test information for subsystems. In the final section the application of AI techniques to layout problems is described. The first chapter in this section discusses EALS (Expert Analogue Layout System), a sophisticated design tool which tackles the problem of enabling demanding, multiple constraint, analogue design problems to be solved by incorporating design expertise as an integral part of the computer package. The final chapter in the book is concerned with the problem of generating optimal channel routing in a gridless router. The objective, in both Chapters 10 and 11, of using the AI approaches discussed, is to alleviate some of the problems, requiring manual intervention, encountered in using algorithmic methods in this part of the design process. Throughout the book the topics described are supplemented with comprehensive lists of references at the end of each chapter.

The editors wish to extend their thanks to the IEE for supporting the production of this volume, to individual authors and, particularly, to Jakki Curr who has collected, collated and corrected material from half a dozen different systems to produce a uniform text.

Gaynor Taylor

Gordon Russell

List of Contributors

Z. M. Alvi
Department of Electronic
  Engineering
University of Bradford,
West Yorkshire BD7 1DP.
United Kingdom.

A. P. Ambler
Department of Electrical
  Engineering
Brunel University
Middlesex UB8 3PH
United Kingdom

B. R. Bannister
Department of Electronic
  Engineering
University of Hull
Cottingham Road
Hull HU6 7RX
United Kingdom

F. P. M. Beenker
Way 4
Philips Research Laboratories
Eindhoven 5600 MD
The Netherlands.

R. G. Bennetts,
Bennetts Associates
Burridge Farm
Burridge
Southampton SO3 7BY
United Kingdom

F. P. Burns
Department of Electrical
  Engineering
University of Newcastle
  upon Tyne NE1 7RU
United Kingdom

M. F. Chowdhury
Department of Electronic
  Systems Engineering
University of Essex
Colchester
Essex CO4 3SQ
United Kingdom

I. D. Dear
Department of Electrical
  Engineering
Brunel University
Middlesex UB8 3PH
United Kingdom

P. B. Denyer
Department of Electrical
  Engineering
University of Edinburgh
The Kings Buildings
West Mains Road
Edinburgh EH9 3JL
United Kingdom

C. Dislis
Department of Eletrical
  Engineering
Brunel University
Uxbridge
Middlesex UB8 8PH
United Kingdom

N. Jeffrey
Department of Electronic
  Engineering
University of Hull
Cottingham Road
Hull HU6 7RX
United Kingdom

D. J. Kinniment
Department of Electrical
  Engineering
University of Newcastle
  upon Tyne NE1 7RU
United Kingdom.

A. M. Koelmans
Computer Laboratory
University of Newcastle
  upon Tyne NE1 7RU
United Kingdom

R. J. Mack
Department of Electronic
  Engineering
University of Essex
Wivenhoe Park
Colchester CO4 3SQ
United Kingdom

R. E. Massara
Department of Electronic
  Systems Engineering
University of Essex
Colchester, Essex CO4 3SQ
United Kingdom

P. J. Miller
Control Techniques
  Process Instruments
Woods Way
Goring-by-the-Sea
Worthing
West Sussex BN12 4TH
United Kingdom

W. R. Moore
Department of Engineering
  Science
University of Oxford
Parks Road
Oxford OX1 3PJ
United Kingdom

J. P. Neil
Department of Electrical
  Engineering
University of Edinburgh
The Kings Building

West Mains Road
Edinburgh EH9 3JL
United Kingdom

C. A. Njinda
Department of Electrical
  Systems
University of Southern
  California
University Park
Los Angeles CA90089
U.S.A.

J. M. Noras
Department of Electronic
  Engineering
University of Bradford
West Yorkshire BD7 1DP
United Kingdom

G. Russell
Department of Electrical
  Engineering
University of Newcastle
  upon Tyne NE1 7RU
United Kingdom

M. F. Sharpe
Department of Electronic
  Engineering
University of Essex
Wivenhoe Park
Colchester CO4 3SQ
United Kingdom

G. E. Taylor
Department of Electronic
  Engineering

University of Hull
Cottingham Road
Hull HU6 7RX
United Kingdom.

K. W. Turnbull
Engineering Leader for
  Synthesis Development
GenRad Design Automation
  Products
Fareham
United Kingdom

Chapter 1
Expert Assistance in Digital Circuit Design

Z. M. Alvi and J. M. Noras

1.1 *Introduction*

This chapter is concerned with the design and analysis of digital systems using IKBS techniques. Present CAD systems are algorithmic in their nature and most are either simulation or layout tools. Moreover, there are a number of problems associated with their learning and use. The user is forced to make early design choices and there is no turning back if any of these decisions are not suitable. The design of digital systems is an ill-structured problem and the use of AI-based design techniques for their design offers several advantages such as ease of use, development, understandability, simplicity and explanation. An expert system called the design assistant (DA) has been developed to demonstrate these advantages.

Since the DA reasons about alternatives without looking into their implementation details it is helpful for the designer in that it is not necessary to make early choices which could have unforeseen effects. The DA is able to design parts of a microprocessor based system and provides facilities for analysing circuits. The use of these techniques can also be used for the design of VLSI systems.

1.2 *Digital Systems*

Digital systems design is essentially a trade-off between conflicting requirements, whether the system is a single custom chip or a combination of a number of custom or proprietary devices. On the one hand the increase in complexity of the systems being designed, the shortage of design time and the need for corrections emphasise the necessity for techniques which can reduce and manage the complexity of the design (Holden, 1988). These techniques include modularisation, standardisation of cells and their composition methods, simple interfaces and automatic tools, all of which affect the silicon area or performance. On the other hand, the area available on the chip is always limited and the pressures of cost and competition may result in compromising designs which have increased coupling capacitance between cells and therefore increased complexity, which, in turn, may result in high risks and longer design times (Russell et al, 1985).

Thus design of digital systems involves finding the right combination of many factors and constraints. With every situation the importance of each of these parameters can vary depending upon the global perspective where each factor may have a certain weighting attached to it. Evidently the size and complexity of such systems, in particular VLSI, makes this process more and more difficult practically.

For example, the requirement may be to design the fastest possible circuit without the cost becoming prohibitively high. It is not readily possible at the start of the design process to calculate exactly the cost without looking at the details of the design at which time it may be necessary to either improve the design or modify the cost limit. This is really an on-going process and may well continue during the life-cycle of the design. Digital design is thus an ill-structured problem and the use of AI techniques can help in designing CAD systems for these kind of design problems (Simon, 1973).

1.3 Present CAD Systems

Many CAD systems are available which help the designer with some of the aspects of design. However, most are either simulation or layout tools with the human designer performing design iterations by hand.

The process of digital systems design, in general, can be divided into a number of design stages (Zimmermann, 1986, Horbst and Wecker, 1986, Rammig, 1986 and Holden, 1987). The four main stages or levels are: the behavioural or algorithmic level, the functional level or the architectural level, the structural or logic level, and the circuit or physical level. As has been pointed out (Russell et. al, 1985), the levels above the functional level are not very clearly defined and sometimes the behavioural level may be replaced by algorithmic level or register transfer level (sequences of data transfer between registers).

High-level synthesis is the design and implementation of a digital circuit from a behavioural description, which generally will not include any reference to the structure of the circuit, and such CAD systems are sometimes called "Silicon Compilers."

In structural compilers the designs are specified as sets of components and their interconnections. Each component is either called from a library or its corresponding component compiler is activated to layout each component on silicon. The final layout is generated by placing and interconnecting the individual components. The designer has to evaluate and, if needed, redesign the layout in order to meet constraints. To help him do so he is provided with analysis tools which can analyse timing, faults, and testability. Structural compilers are being used for the development of application specific integrated circuits (ASICs). These are integrated circuits which, as the name implies, have a special purpose or application and can not be used for any other application. Examples of these are special ICs used in military applications, special purpose hardware systems such as telephones etc.

In behavioural compilers the designs are specified by sets of input and output ports, a behaviour of output ports in terms of the input ports, and sets of constraints that the design is required to meet. There are two types of behavioural compilers. The first type integrates all the tools described above and automatically generates layout directly from input. The second type uses synthesis as a preprocessor for a structural compiler.

Most of these systems are geared towards simulation and layout. They are based on conventional programming languages which encourage algorithmic thinking. Although they are quite powerful in that they provide extensive facilities for designing, testing and validation they are mostly based on algorithms. In other words they use a pre-determined strategy for their designs. This algorithmic approach to design may be quite useful for a small set of design problems, however, the kind of design manipulations that the human designers have to do are not generalised enough to be captured into algorithms. Consequently, for these kind of problems the present CAD systems do not perform as well as human experts.

Using these tools the designer has to make design choices very early in the design process. The problem with this is that the designer is unable to foresee the long term consequences of these choices. By the time he realises the effects of these decisions it is generally too late to go back and change the decisions. This kind of backtracking is a very important part of design but is not included in any of these systems.

If it is possible to evaluate the long term effects of design decisions at an early stage it can be quite helpful for the designer and can, therefore, result in better quality designs. However, design iterations are performed by hand. Moreover, being algorithmic in their nature, these systems do not accept existing designs as a starting point of design (Dasgupta, 1989).

Also, they do not provide any information about their design decisions and indeed about their design processes, and any one who may be interested in monitoring the design of digital systems by using this kind of software will not be able to do so. There is simply no way of following what goes on behind the scenes. They do provide help to the designer but are unable to enhance his designing ability.

Moreover, as the design objects become more complicated and sophisticated, designers need to rely more on computers for processing such information. There is therefore a need for intelligent CAD systems (Gero 1985 and 1987 and ten Hagen,1987) which can help improve the designer's abilities by providing him with information about the design process. There does not, however, appear to be any clear definition of so-called intelligent CAD systems.

1.4 The Nature of Design

A characteristic of design problems is that the requirements are generally incomplete and imprecise. The problem of incomplete requirements is generally associated with human characteristics and has been discussed by Simon (1976,1981 and 1982). Usually the requirements are extended or improved, perhaps many times, during the course of the design.

Having finished a design which may be an artifact or abstract definition the next stage is the testing or verification of the design against the requirements which provided the initial stimulus. This process determines whether the product of design serves the purpose for which it was designed or needs further modification. The designer may have to go back to the earlier stages and change some or all of the parameters and come up with a modified or improved design. The modified design goes through another process of evaluation and testing and the process goes on. Existing CAD tools do not allow for this complex cycling.

Consider a situation in which the user who requires a certain product of design, whether it is an artifact or an abstract plan, is not entirely clear about what he wants. In other words he knows some of his requirements and is not entirely sure whether they are complete. Indeed, it may be the case that a design project begins with one or a small set of requirements that collectively represent the basic problem and enable the project to be defined. These are then merely the "top-level" or the most fundamental requirements, there may indeed be more as they unfold. As stated by Dasgupta (1989), problem identification in many design situations may be quite incomplete or poorly understood at the time the design begins.

In these conditions the designer may start the design process using the initial or the first set of requirements which the user provides. However, since the user is not entirely sure about his or her requirements, he either changes or modifies some of them later. If the designer has reached a stage where it is possible for him to incorporate the changes he may be able to do so without much difficulty. If such is not the case then the new or modified requirements may well have to be overlooked under the circumstances.

To design according to the revised or modified requirements the designer may have to either start from scratch or change some of his design decisions. The evaluation and testing process will need to be repeated again.

As a consequence of the above, there are two important implications Dasgupta (1989):

1) Since the requirements are incomplete, it is difficult to distinguish between "requirements" and "design". In other words, those requirements which only become clear as the design proceeds, and are as such part of the requirements, may be confused with a part of design itself.

2) Irrespective of (1) above, it is obvious that in reasonably complex design problems, the clarification, development or expansion of the given set of requirements is an inherent component of the design process.

1.5 Design is a "Satisficing" process

Generally, the design process involves making decisions which are typically of the form:

1) If there are a number of interacting objectives or goals, how to arrange them in terms of priorities. In other words, if the interactions between different constraints are only known in abstract terms, which constraints are to be handled before others and which ones are to be ignored during the initial stages of the design process.

2) How to choose among different choices that are equivalent, say, in terms of functionality, performance, and/or cost, for example, the components of a circuit.

Since most design problems are complex and the designer has imperfect knowledge of the long-term consequences of design decisions, the rationality that the designer can bring to a design problem is limited. This concept of "bounded rationality" led Simon (1969) to suggest that for design problems of any reasonable complexity, one must be content with "good" rather than "best" solutions. That is, the designer sets some criterion of "satisfactoriness" for the design problem and if the design meets the criterion, the design is considered to be solved, even if temporarily. He therefore states that design problem solving, and indeed many other types of problem solving, are "satisficing" procedures in that they produce satisfactory rather than optimal solutions.

1.6 The Artificial Intelligence Paradigm

As stated, the AI paradigm addresses some of the problems other paradigms do not handle very well, in particular ill-structured problems (Simon, 1973). It views design as search through a problem space containing semi or partial solutions, the starting point of which is the initial state. Each design stage or step can be represented as a potential solution in the problem space. Rules or operators are used to move from one design stage to the other. The search is guided by heuristics (Feigenbaum *et al*, 1963). Using this approach, design problem solving can be represented as a state space search (Newell and Simon, 1972). The result of applying a sequence of legal operators is to move through the problem space (Langley *et al*, 1987). According to Harmon and King (1985) problem solving is the process of starting in an initial state and searching through a problem space in order to identify the sequence of operations or actions that will lead to a desired goal.

The characteristics of ill-structured problems can be described as follows (Cross, 1989):

1) There is no systematic definition of the problem. During the initial stages of design, the goals are generally vague and many constraints and criteria are unknown. The context of the problem is quite complex and messy, and difficult to understand. As the design proceeds, temporary definitions of the problem are set, but these are unstable and subject to changes as more information about the problem is gathered.

2) Any definition of the problem may be inconsistent due to either incomplete initial specifications and or due to the nature of the problem itself. Many inconsistencies may have to be resolved to solve the problem. Generally, as the problem is solved more inconsistencies come to light.

3) Definitions of the problem are solution-dependent. In other words, the ways of defining the problem are dependent upon ways of solving it; it is difficult to define a problem without

specifically or otherwise referring to a solution concept. The concept of the solution influences the concept of the problem.

4) Solution proposals are used to understand the problem. Many assumptions and uncertainties about the problem can be highlighted only by proposing solution concepts. Many constraints and criteria emerge as a result of evaluating solution proposals.

5) There is no single unique solution to the problem. There may be a number of different solutions which are equally valid responses to the initial problem. Moreover, the solutions cannot be evaluated objectively as either true-or-false or right-or-wrong, rather they are compared and assessed as good or better, appropriate or inappropriate.

6) Generally there are time constraints which require that the problem be solved in a certain amount of time.

As a consequence of these characteristics the combinations of the states and the operators grow exponentially leading to an unbounded search or problem space. The process of searching an unbounded problem space can take a large amount of time thereby leading to time constraint violation. Though it could be argued that in the ideal case it would be possible to try out all the possibilities that may exist to reach a solution, in other words conducting an exhaustive search, or the brute-force method according to Feigenbaum (Feigenbaum *et al*, 1963). Clearly, because of the kind of problems associated with digital design, it can be called an ill-structured problem.

Generally it may not be possible to conduct an exhaustive search and arrive at the best solution within reasonable time. There may be too many constraints and, for ill-structured problems, the time required to find the best solution may not be permissible. To cut down on the numerous possibilities that may exist in an unbounded search space the designer may have to make certain assumptions or ignore some of the relatively unimportant parameters.

Given that the designer is an expert in his or her domain, reducing the number of constraints or variables can mean an overall reduction in the quality of design. In other words, the above process may diminish the quality of the design and the solution to the problem may not be the optimal.

When experts are faced with ill-structured problems they use their experiences, rules of thumb, and other guidelines which help them in pruning the search space to reduce it down to a manageable size. These are called heuristics and the AI paradigm incorporates these as part of the human problem solving strategies. According to Feigenbaum (1978) the use of knowledge based methods are considered to be most appropriate for implementing heuristics on a computer. The implementation of the design knowledge in the DA is based on these principles.

There is a major trend towards using knowledge-based expert systems for the design of VLSI system. This is a direct result of the increasing complexity of such systems. The use of the IKBS approach promises to alleviate some of the problems of VLSI designs by being able to "imitate" human designers (Fujita *et al*, 1986, Mitchell et al, 1985A, Mokhoff, 1985 and Rosensteil *et al*, 1986).

A number of expert systems have been developed mostly in the research community for designing different aspects of digital systems and indeed VLSI. There are hundreds of thousands of transistors on a single chip. This level of complexity has resulted in tremendous amount of detail to be handled and is proving to be a major limitation in achieving cost-effective, low-volume, special purpose VLSI systems (Kowalski *et al*, 1985).

According to Kowalski because of the combinatorial explosion of details and constraints the problem of choosing the right kind of architecture or implementation "does not lend itself to a recipe-like solution. However, Knowledge Based Expert Systems (KBES)provide a framework for just such problems that can be solved only by experts using large amounts of domain-specific knowledge to focus on specific design details and constraints". Moreover, as VLSI designers remain scarce it is even more important that instead of starting each new project from scratch, an existing design be modified to cater for new requirements.

Current research in the KBES field relating to VLSI design is proceeding along two parallel paths. Firstly, the synthesis of integrated circuit chips, and secondly the analysis of already developed designs (Mitchell *et al*, 1985A, Mokhoff 85 and Rosensteil *et al*, 1986). The analysis of finished designs appears to be relatively easy as compared to designing new ones, since it requires matching the final design with the original specification, and the required information is already available in the form of specification. Synthesis, however, requires that the knowledge base must first be constructed. Consequently the main thrust of research activity has been in the former kind of problems, i.e. analysis of existing designs.

According to McDermott (McDermott, 1981), there are two additional reasons why KBES systems are specially suited for application to design systems:

1) Since in a design problem there are a large number of possible solutions from which the optimal one must be selected, the KBES approach can reduce the number of possible solutions using its knowledge.

2) Most designs require a large number of constraints to be handled and these constraints may change from one situation to another: such problems are beyond the capabilities of algorithmic solutions.

1.7 *The Design Assistant*

This section presents the details of implementation of the expert hardware design system, called the design assistant (DA). The first part discusses the general concept of design automation systems, and the aims and requirements set forth for the DA. The second part briefly describes the structure of the DA.

The DA is a hardware design system, which uses a rule based approach for the design and analysis of digital hardware in general, and microprocessor based hardware in particular. The design of hardware, at present, is limited to the design of the memory system for a given set of memory blocks from the user. The design uses trade-offs for comparing choices amongst the components that can be used based on factors like speed, chip count, area, and cost. It is also able to perform circuit analysis by simulating digital circuits for logic and evaluating the timing behaviour.

The two wings of the DA, i.e., design and analysis, have a separate interface to the user and at present run as two different applications.

1.7.1 General scheme of a design automation system

Design automation systems can be classified into two main categories; fully autonomous systems and design assistants. The fully autonomous systems are ones which do not interact with the user and are difficult to adapt to changes in the domain while the design assistants provide more flexibility from the point of view of both adaptability to changes in the domain and the interaction with the user. McCullough (1988) suggests that it is more useful to develop design assistants rather than the other type.

A general scheme for a design automation system for digital design is given in Figure 1. It represents the stages of a design system which starts with requirements capture from the user and the result is a hardware design which satisfies these requirements. The first block marked "user requirements" represents the interaction between the design automation system and the user who supplies requirements to the system: before an attempt is made to design a target system, they are analysed for consistency and completeness.

As has been discussed earlier, initial requirements are mostly incomplete, imprecise or inconsistent. For example, the user may ask for a hardware system which can communicate over a communication network without specifying the speed of the network, or may ask for a real-time system without working out the exact timing responses that the target system is required to satisfy.

12 Expert assistance in digital circuit design

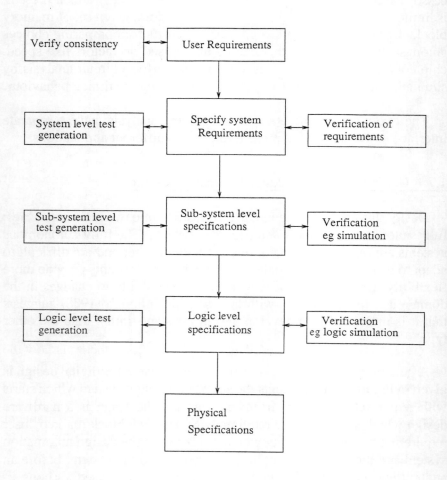

Figure 1 Digital design automation system
(General Scheme)

Therefore, before a set of specifications for a hardware design is made it is important to analyse the requirements. Once this process in completed the next stage is the formulation of requirements for the hardware that is to be built. This is represented by the block marked 'specify system requirements'. This stage is the highest level of system design and the designer is concerned with the analysis and type of data to be handled, the analysis of performance in terms of the throughput of the desired system and evaluation of alternative choices from the point of view of feasibility related to speed, power, cost, area, etc.

At this level, and in fact at all the levels, there is in principle, a test generator to ascertain the correct working of the design (shown by the blocks on the left). In addition to this there is an additional requirement that the design be verified at each level. This is shown by the blocks on the right. The last level shown is the physical level i.e. the implementation level and represents the physical layout on a PCB or an integrated circuit.

At the level of system design the designer is concerned with the overall structure of the system, for example, in the case of a microprocessor based system the designer is mainly concerned with the main blocks which are the CPU, the memory and the input/output. Also the concern is how the memory is to be arranged, i.e., for multipurpose systems where there is to be more than one user at a time the memory has to be partitioned into blocks of different sizes which represents the block size allocated to each user. Certain areas of memory are only accessible to the system manager and there is a supervisory mode of the CPU which permits this kind of operation.

In the case of input/output the main concern is the number of external devices to be handled, the system response time to the external interrupts, how to prioritise the interrupts if more than one device tries to communicate with the processor etc. Obviously, at this level the designer is not concerned with how but what to implement.

The DA fits into this general scheme of design. Presently, the scope has been narrowed onto microprocessor based systems. The DA is an expert system for the design of these systems given an initial set of top-level requirements. These requirements are essentially specifications for a digital system containing a microprocessor at its heart. The user is not required to specify the details of the system required, rather the idea is to make a design assistant for the use of people who are required to design such hardware but are either not familiar or do not have the time to program the multitude of CAD programs available.

Although there seem to be all sorts of CAD programs available, there is hardly any system which can perform design and analysis of a digital system given a top-level set of requirements. As stated earlier, most of these systems are tools for simulating or analysing such systems. With the advent of the VLSI era the focus of these programs has been on the design or the layout of integrated circuits. Other programs are used as tools for testing, analysing, and verification of these designs. The DA is an attempt to alleviate some of these problems by helping the designer to think and reason about high level design decisions without considering their implementation aspects. It is also potentially possible to evaluate the long term consequences of different design choices and make decisions by evaluating these effects. Furthermore, it may also be possible to build a library of existing designs and either re-use them as they are or slightly modify them for use.

1.7.2 Aims of the DA

The DA is intended to assist with hardware development and can produce specifications for single board CPU based systems when supplied with details of the requirements for memory and I/O devices. It is expected to help produce hardware designs without the user needing to learn the principles of digital systems designs, indeed without learning to use the various CAD tools available. In addition to this it is also expected to analyse the designs by performing simulation, timing analysis, and also interfacing to an existing PCB/VLSI layout package.

The design of microprocessor based systems is an ill-defined, complex problem. The DA is therefore required to use IKBS techniques for the design of hardware. At the same time it is important that the design process be made easy to comprehend for the users. This requirement is useful if the DA is to be used for explaining the design to the users, for their education or to give confidence in the design.

The analysis of circuits involves the testing and verification that the circuit does indeed work as required. This may involve the analysis of the truth-table of a circuit. In the case of memory design the truth-table may be used to verify that the appropriate address lines do indeed select a specified part of memory or any other device such as an input/output port.

The analysis of logic is used to work out the functional behaviour of a circuit. The functional behaviour represents the output responses of a circuit for each set of inputs. If there are multiple inputs and multiple outputs, it is important to be able to verify that for each set of possible inputs there is a set of specified outputs. The analysis of timing is another part of circuit analysis which is mainly concerned with the response times of a circuit given a change at its inputs.

The analysis of timing is carried out after the logic analysis has been performed. There are two possible scenarios for the use of timing analysis. In the first scenario the objective is to find out whether the analysed logic of the designed circuit performs its logical function at a desired speed or in a given amount of time. This problem arises due to the fact that even very tiny circuits, e.g., the transistors in integrated circuits, have a non-zero delay before their outputs respond to changes in their inputs.

The second scenario is used to compare the timing responses or propagation delays of two circuits which have similar functional capabilities. Obviously if one circuit has a faster response time or smaller propagation delay it can perform the same logical function faster than the other circuit.

In large circuits there may be many sub-circuits, resulting in many different paths between an input to the circuit and its output. If it is possible to identify these paths then it is also possible to find out which path or paths have the lowest propagation delay and which have the highest. If the critical paths can be identified and analysed, the task of the circuit designer can be greatly simplified. If the propagation time of the sub-circuits in the critical paths can be reduced it can improve the propagation time of the entire circuit.

Finally the DA is required to provide an interface to a PCB/VLSI layout package so that the analysed design can be implemented on hardware.

The specifications provided to the DA, as stated earlier, can be very general specifying only the nature of the application like real-time control, digital signal processing etc. The DA is to design the hardware for these requirements without going into the internal details of the hardware components that it uses. In other words the DA analyses the functioning of the components or subcomponents not in terms of their internal structure, rather in terms of their external characteristics.

For example the functional behaviour of a full adder is independent of its internal structure, it may be made up of random logic using primitive gates or it may be made up of half adders. It is not intended to look beyond the functional behaviour of circuits or sub-circuits. It is only concerned with external features like the speed, area, power consumption and cost of these circuits. These factors are important from the point of view of system design and play an important role in the overall design and structure and life cycle of a design. A design which is cheap and robust is preferable to one which is fast but unreliable or costly. The DA aims to design systems using these factors rather than using the technical features of the components like the internal setup and hold times etc.

It is intended that successful implementation of expert system techniques to this problem will be extended to the compilation of digital VLSI systems from libraries of circuit elements.

1.7.3 Requirements for the Design Assistant

Based on the aims specified above, the DA is required to perform some or all of the following functions:

1) Be able to capture requirements from the user in an interactive and friendly manner so that the user finds it easy to talk to the system. These requirements generally can be non-technical in nature, for example, design a stand-alone microprocessor based system which is large enough to manage programs of length not more than 64K bytes.

2) Analyse the requirements for consistency, completeness and preciseness.

3) Determine whether a hardware system can in fact be built which can perform the required function. Also whether a microprocessor based system is really needed.

4) Design the required hardware using intelligent techniques.

5) Use either commercially available off-the-shelf integrated circuits to design the hardware or use an ASIC library of components. The advantage of using the commercial hardware is that there is a tremendous variety of components available including VLSI components such as microprocessors, and in the early stages of its implementation it will be helpful. The disadvantage with ASIC library components is that they are not widely accepted.

6) Provide a platform for interface to simulation and layout packages so that the design can be tested for different parameters. The interface platform can be quite simple, for example to write an input file. For hardware designs to be implemented on integrated circuits it may be better to use existing software rather than write new software.

7) As a part of the general design process the output from the DA is to be in such a form that it can be tested for consistency and correctness not just for logic but also for timing. This can be performed, for example, by simulating the designed hardware.

8) Be simple to operate, maintain and update. Mostly the people who write software do not have to maintain it later. This is particularly true of large software systems and has been felt quite strongly in the case of the XCON system.

9) Finally, as an expert system it should provide a framework for explaining its reasoning i.e. be able to justify its design decisions so that the design process can be analysed for any inconsistencies or errors and can thus be improved or modified. The explanation of the design decisions is a requirement associated with intelligent systems.

1.7.4 Present status

The DA has been implemented in the Automated Reasoning Tool (ART) (ART, 1985), and can perform both design and analysis using knowledge based system techniques. The design is, at present, limited to the design of memory for a microprocessor based system using off-the-shelf memory chips. In principle, any microprocessor can be specified, the present version, however, can only design for the Motorola MC68000 microprocessor. The output is a list of these chips, their interconnections, the decoders for selecting different memory chips corresponding to different addresses generated by the CPU, decoders for selecting different blocks of memory like ROM, EPROM, RAM etc., how these are connected and their start addresses in terms of the address lines selecting a block.

In memory design the DA uses trade-offs for devices in terms of speed, chip count, cost, area. There is no explicit representation of the upper bounds for these constraints at present. The DA designs the memory using the chips available and makes the trade-offs using the above constraints.

For a given circuit, the DA can analyse the circuit which includes logic analysis such as the truth-table, finding the output values of nodes in the circuit asked for by the user, find the delays at different nodes in the circuit, find the maximum delay at any node (measured from an input node), find a link between an input node and a node specified by the user.

The last function is particularly useful in the design of VLSI where it is important to know which inputs control which outputs. The interface to both wings of the DA is separate: keyboard input in the case of design and menu driven for analysis.

1.7.5 Functional arrangement of the Design Assistant

The structure of the design assistant is shown in Figure 2. Functionally the DA can be divided into three parts; the user interface, the main body or the intelligent knowledge base, and the PCB/VLSI interface. The work reported here is mainly concerned with the development of the main body or the design and analysis part. It is expected that once this is finished then the work can be extended to include the intelligent interface to the user and the back end to the layout environments.

1.7.6 The knowledge base

Rules are used to represent the domain-dependent design knowledge in the DA and are the main reasoning mechanism. Additionally, data and design information both are represented using Object-Oriented Programming (OOP). By combining the rule based inference mechanism and Object-Oriented Programming it has become possible to make the DA simple, easy to develop, maintain, update and understand.

For example, to keep the knowledge base small and manageable only the essential information about the objects is specified at the time of adding knowledge to the DA, such as the addressing-range of memory components. Only when this value is required by the DA during run-time is it calculated and used by the DA. When the number of data is very large this is quite helpful in maintaining the knowledge base.

The use of OOP has also been used for designing the memory subsystem and only two main objects are used to replicate the address decoding table which essentially contains all the relevant information necessary to design a microprocessor based system. The advantage is that it is quite

20 Expert assistance in digital circuit design

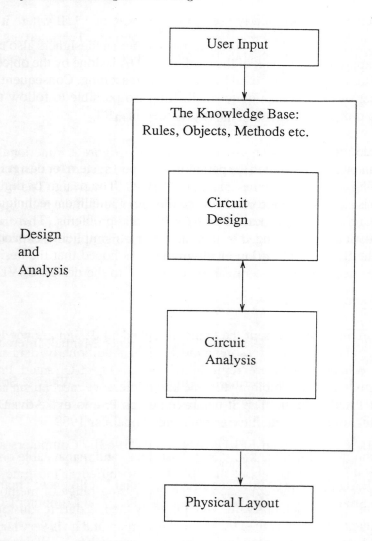

Figure 2 Functional view of the design assistant

easy to understand the design process since the human designers also use the same approach. The bulk of the work in the DA is done by the objects and the rules are essentially used for meta-level reasoning. Consequently, the number of rules is kept quite small. It is also possible to follow the reasoning process using the facilities provided in ART.

1.8 Conclusion

This chapter has presented the details of an expert system for designing digital systems, called the design assistant (DA). The design of digital hardware is an ill-structured problem and the use of intelligent techniques helps in automating the design process of such problems. There are several advantages of using this approach such as simplicity, expandability, ease of use, explanation of reasoning. It is hoped that the design techniques used in this system can be expanded to the design of VLSI systems.

1.9 References

ART (1985). Automated Reasoning Tool, Reference Manual, Inference Corporation, USA.

CROSS N., (1989), "Engineering Design Methods", John Wiley & Sons.

DASGUPTA, S. (1989). "The Structure of Design Processes", Advances in Computers vol.28, M.C. Yovits ed., Academic Press 1989.

FEIGENBAUM, E.A. and FELDMAN, J. ed. (1963). "Computers and thought", McGraw Hill, San Francisco, USA.

FEIGENBAUM, E.A. (1978). "The Art of Artificial Intelligence: Themes and Case Studies of Knowledge Engineering", Proc Nat Computer Conf (AFIPS) pp. 227-340.

FUJITA, T. et al (1986). "Artificial Intelligent Approach to VLSI design." in Design Methodologies, S. Goto (Ed.) Elsevier-Science Pub.

GERO, J.S. ed. (1987). "Expert Systems in Computer Aided Design", Proc IFIP WG 5.2 Workshop in 1987, Sydney, Australia, North-Holland, Amsterdam.

GERO, J.S. ed. (1985). "Knowledge Engineering in Computer Aided Design", Proc IFIP WG 5.2 Workshop in 1984, Budapest, Hungary, North-Holland, Amsterdam.

HARMON, P. and KING, D. (1985). "Artificial Intelligence in Business: Expert Systems", Wiley Press 1985.

HOLDEN, A. (1988). "An Intelligent Assistant for the Interactive Design of Complex VLSI systems". 2nd IFIP Workshop, Cambridge England.

HOLDEN, A. (1987). "Knowledge based CAD for Microelectronics." North Holland.

HORBST, E. and WECKER, T. (1986). "Introduction: The Evolution of the Logic Design Process." in Logic design and simulation, E. Horbst (ed.), North-Holland (Elsevier-Science Pub), The Netherlands.

KOWALSKI, T.J., (1985) An Artificial Intelligence Approach to VLSI Design, kluwer Academic Pub, Boston USA.

LANGLEY, P. et al. (1987). "Scientific Discovery: Computational Exploration of the Creative Processes", MIT Press 1987.

McCULLOUGH, J. (1988). Position Paper on Intelligent CAD Issues, Proc IFIP WG 5.2 Workshop on Intelligent CAD, Cambridge UK.

MCDERMOTT, J., (1981) Domain Knowledge and the Design Process, 18th Design Automation Conference.

MITCHELL *et al*, (1985) T.M. Mitchell, L.I. Steinberg, J.S. Shulman, "A Knowledge-based Approach to Design", IEEE Trans Pattern Analysis & Machine Intelligence, vol PAMI-7 no 5, Sept 1985.

MOKHOFF, N. (1985). "AI Techniques Aim to Ease VLSI design." Computer Design, March.

NEWELL A. and SIMON H.A., (1972) "Human Problem Solving", Prentice-Hall, Englewood Cliffs, N.J., USA.

RAMMIG, F. J. (1986). "Mixed Level Modelling and Simulation and of VLSI Systems", in Logic Design and Simulation, E. Horbst (ed.), North-Holland (Elsevier-Science Pub), The Netherlands.

ROSENSTEIL, W. and BERGSTRASSER, T. (1986). "Artificial Intelligence for Logic Design", in Logic design and simulation, E. Horbst (ed.), North-Holland (Elsevier-Science Pub), The Netherlands.

RUSSELL, G. *et al*, (1985). "CAD for VLSI", van Nostrand Reinhold, England.

SIMON, H. A. (1969). "Sciences of the artificial". MIT Press USA.

SIMON, H. A. (1973). "The Structure of Ill Structured Problems". Artificial Intelligence vol 4.

SIMON, H. A. (1976). "Adiminstrative Behaviour" (3rd edition). The Free Press New York.

SIMON, H. A.(1981). "Sciences of the Artificial" (2nd edition). MIT Press, Cambridge, Mass, USA.

SIMON, H. A. (1982). "Models of Bounded Rationality" vol 2. MIT Press, Cambridge, Mass, USA.

ten HAGEN, P.J.W. and TOMIYAMA, T. eds. (1987). "Intelligent CAD Systems 1 Theoretical and Methodological Aspects", Springer Verlag, Berlin.

ZIMMERMANN, G. (1986). "Top-down Design of Digital Systems," in Logic Design and Simulation, E. Horbst (ed.), North-Holland (Elsevier-Science Pub), The Netherlands.

Chapter 2
Use of a Theorem Prover for Transformational Synthesis

A.M. Koelmans, F. P. Burns and D. J. Kinniment

2.1 Introduction

The use of formal methods in VLSI design is currently a very active research area. This requires the specification of the behaviour of a hardware design in a formal, mathematically rigorous manner. Such specifications can subsequently be manipulated by proof systems (usually called 'proof assistants' or 'theorem provers') to prove the equivalence of hierarchical or temporal properties of hardware designs.

The development of hardware synthesis systems seems to have taken place without the involvement of formal methods. The problem in hardware synthesis is to take an initial specification, frequently in the form of boolean equations or register transfer statements, and map these efficiently into silicon. Such languages do not lend themselves very well to formal reasoning, and recent research efforts have only concentrated on mapping formal languages into silicon. The use of formal languages makes it difficult to generate an efficient layout, since a much larger design space needs to be explored than when low level languages are used. This means that formal specifications will usually need to be transformed, in several stages, to a low level form which can then be mapped into silicon. A major problem here is to ensure that the applied transformations do not invalidate the original specification. This, of course, should involve the use of a proof system, and this paper describes our efforts in this area.

We describe a prototype tool which integrates a theorem prover into the design environment in such a way as to ensure functional and transformational correctness at all times. The system is driven from a hardware description language developed at Newcastle called STRICT, and it is interactive. Our aim in developing this tool is to make it easier for the designer to ensure correctness of the final implementation, without losing the benefit of his skill and experience.

There are two proof systems which have already shown great promise in the area of hardware design: HOL (Gordon, 1987) and the Boyer Moore theorem prover (Boyer and Moore, 1988). Both use a specification language based on formal semantics, have been used to verify substantial hardware designs, are widely used in the VLSI research community, and are readily available. HOL can be most accurately described as a proof assistant, that is, it requires interactive commands from a user who is an expert in the internal workings of the system. The Boyer Moore theorem prover requires all its input from a text file, and runs without interactive user input during the proof process. For this reason, we felt that this prover is a more suitable tool in a hardware synthesis environment, and we therefore decided to use it in conjuction with our synthesis tool.

The paper is organised as follows. In section 2.2 we introduce the basic features of high level synthesis systems. In section 2.3 we describe the Boyer Moore theorem prover, and in section 2.4 an example of its use is presented. In section 2.5 we describe the basic features of our tool, and we finally present an example of its operation in section 2.6.

2.2 *Transformational synthesis*

An excellent introduction to high level synthesis systems can be found in the overview paper by McFarland (McFarland *et al,* 1988). All high level synthesis systems initially operate upon a user specified behavioural description in an appropriate high level language. The language we use for this purpose is called STRICT (Campbell *et al,* 1985), which is a conventional hardware description language which allows specifications to be written in a functional notation. This description is parsed into an internal format. This is followed by the scheduling and allocation phase, during which the basic functional units of the design are determined and the basic hardware units are assigned to these functional units, together with memory elements and communication paths. The resulting design is then fed into conventional floorplanning and routing tools to produce the final chip layout. State of the art examples of high level synthesis systems are Cathedral (Deman et al, 1986) and the Yorktown Silicon Compiler (Brayton et al, 1988).

The high level synthesis community is split into two camps: those who want to fully automate the design process, and those who want to make use of the designer's experience by providing interactive input during the design process. In fully automatic systems, much of the current reseach is focussed on the provision of algorithms to ensure that reasonably efficient hardware is generated from a large number of possible implementations.The interactive approach encourages designers to investigate different designs through experimenting with different architectures generated from the same specification by, for instance, serialising operations upon the same device or by allocating many operations to devices operating in parallel. We firmly believe in the interactive approach, and the tool described in this paper was inspired by SAGE (Denyer, 1989), a sophisticated interactive synthesis system, but one that does not incorporate verification features.

A specification of an algorithm in a high level language will in general be unsuitable for direct translation into silicon. It will be necessary to perform transformations which preserve the behaviour, but generate a much more compact silicon implementation (McFarland et al, 1988). The problem is then to ensure that the applied transformations preserve the behaviour (Camposano, 1988), and in general to verify the correctness of the ultimate design produced by the system against the initial specification. Research in this area is reported in (Verkest et al, 1988, Elmasry and Buset, 1989 and Finn et al, 1989). As mentioned in the previous section, we use the Boyer Moore theorem prover to achieve this goal.

2.3 The Boyer-Moore Theorem Prover

The Boyer Moore theorem prover (Boyer and Moore, 1988) was initially developed at the University of Austin as a tool for research in Artificial Intelligence applications. It has been successfully applied in such diverse areas as list processing, number theory, protocols, real time control, and concurrency. The largest proof to date performed by the prover is that of Goedel's Incompleteness Theorem (it is distributed with the source code of the prover). It was used by Hunt (Hunt, 1985) to verify the correctness of a microprocessor design, thereby confirming that this prover could be used in the VLSI design area. The prover is written in Common Lisp, and is distributed with a specially modified version of the Kyoto Common Lisp interpreter. The software runs on a wide range of machines.

The Boyer Moore logic is a quantifier free, first order logic with equality and function symbols. The input language to the prover is a variant of Pure Lisp. This input language is used to present the prover with a sequence of so called events. The most important events are: definitions of (often recursive) functions to be used in proofs, and theorems. The keyword DEFN starts the description of a function; all functions must conform to certain rules with respect to termination, which are checked by the prover before accepting the function definition. If a function calls other functions, these must have been previously defined. The keyword PROVE-LEMMA initiates a description of a theorem. A theorem will be acceptable if it can be proven by the prover. The prover will attempt to do this using its built-in logic inference rules, definitions, and previously proved theorems. It will in general be necessary to build a database of carefully chosen and carefully ordered theorems to perform complex proofs.

As an example, we can define a function DOUBLE which doubles an integer recursively, and we then prove that this is the same as multiplying by 2:

```
(defn double (x)
  (If (zerop x) 0
      (ADD1 (double (SUB1 x))))
  )
)

(prove-lemma double-is-times-2 (rewrite)
  (implies (numberp x)
    (equal (double x) (times 2 x))
  )
)
```

The theorem includes the hypothesis that x must be a number, and requires it to be stored as a rewrite rule. After it has been proved, the prover will subsequently replace occurences of (double x) by (times 2 x).

The prover has a number of powerful built-in heuristics, one of which is the capability of performing mathematical induction. This means it can cope most efficiently with recursively defined functions (which are not necessarily the computationally most efficient). The prover is also distributed with a number of databases of useful functions and theorems which it frequently will use in proofs. These can be loaded at the request of the user.

The prover has a facility whereby the user can define his own abstract data types. This is usually referred to as the Shell principle. For the purpose of hardware verification, Hunt (Hunt, 1985) has added a data type for use with bit vectors, and a whole set of associated theorems. These can be loaded automatically when the prover is started.

If the prover fails to complete a proof, this will not necessarily mean that there is something wrong with the theorem. It may be that if the proof is rather complex, the prover will not 'see' the right path through the proof. In this case it will be necessary to 'educate' the prover by first proving smaller theorems which prove parts of the big proof, so that the prover will subsequently use these to complete the big proof. In many cases these smaller theorems will be useful in themselves, and can be added to the users' database if he wishes. The prover will therefore frequently need help from the user, which means that in a sense it is not fully automatic.

The prove-lemma construct can have as an optional parameter a list of hints, which can tell the prover to use certain theorems with certain parameters, ignore other theorems, or perform inductions in a specified manner. Some of these features are shown in the example which now follows.

2.4 A Proof Example

We want to prove that for all positive n the integer quotient of n*x and n*y is equal to the integer quotient of x and y, ie. we will try to prove that

(equal (quotient (times n x)
 (times n y))
 (quotient x
 y))

Such a theorem occurs frequently when trying to prove properties of bit vectors, with n having the value 2. We will call this theorem QUOTIENT-TIMES-TIMES. Trying to prove it without the aid of additional theorems fails - the prover tries to solve the problem by using induction, which is not the right approach.

All proof times mentioned below are as measured by the prover on a Sparcstation. A substantial portion of Hunt's database was loaded first. The recursive definition of QUOTIENT (Boyer and Moore, 1988) is:

(defn quotient (i j)
 (if (equal j 0) 0
 (if (lessp i j) 0
 (add1 (quotient (difference (i j) j)))

That is, the quotient is performed by repeated subtraction, while counting the number of subtractions performed.

The proof can be split up in three cases: x<y, x == y and x>y. The first case is described by the following lemma:

(prove-lemma q1 (rewrite)
 (implies (and (greaterp y x)
 (greaterp n 0))
)
 (equal (quotient (times n x) (times n y))
 (quotient x y))

)
)
)

The prover's builtin arithmetic rules are not sophisticated enough to cope with this, so we first prove another lemma to tell it that if x<y then n*x<n*y:

(prove-lemma q1-first (rewrite)
   (implies   (and (greaterp y x)
                  (greaterp n 0)
)
     (greaterp (times n y) (times n x))
)
)

The proof takes 6 seconds. Now lemma q1 is proved in 9 seconds, using q1-first and the prover's builtin rules. The case x == y is easily proved from builtin rules. It takes just two seconds:

(prove-lemma q2   (rewrite)
   (implies   (and (equal y x)
                  (greaterp n 0)
)
   (equal   (quotient (times n x) (times n y))
          (quotient x y)
)
)

The third case, x>y, is the most general one and fails if tried on its own. It can be proven if one realises that if x>y, then x = a*y + b, where a>0, and b<y. Thus we first try the following lemma:

(prove-lemma q3-first (rewrite)
   (implies   (and (greaterp a 0)
                  (greaterp y b)

```
                    (greaterp n 0)
                    (equal (plus (times a y) b) x)
                )
        (equal   (quotient (times n x) (times n y))
                 (quotient x y)
        )
    )
)
```

The proof of this lemma is entirely achieved by simplification using various database rules, and takes 7 seconds.

In order to prove the lemma for the case x>y we need to tell the prover to use this lemma, while substituting (quotient x y) for a, and (remainder x y) for b. This will allow the prover to check that the hypothesis (greaterp x y) holds. Note the use of a hint in this lemma:

```
(prove-lemma q3    (rewrite)
    (implies    (and (greaterp x y)
                     (greaterp n 0)
                )
        (equal   (quotient (times n x) (times n y))
                 (quotient x y)
        )
    )
    ((use    (q3-first    (a (quotient x y))
                          (b (remainder x y))
             )
     )
    )
)
```

The proof takes 40 seconds. The main lemma still cannot be proven - we have to tell the prover that the previous three cases together constitute any possible combination of x and y:

```
(prove-lemma q-bridge (rewrite)
    (implies    (and (greaterp n 0)
```

             (or   (greaterp y x)
                    (equal y x)
                    (greaterp x y)
)
)
       (equal (quotient (times n x) (times n y))
               (quotient x y)
)
)
 ((use (q1))
  (use (q2))
  (use (q3))
)
)

This proof takes 6 seconds. We are now ready to do the final proof, which takes 2 seconds.

(prove-lemma quotient-times-times (rewrite)
    (implies (and (greaterp n 0)
                    (numberp x)
                    (numberp y)
)
        (equal (quotient (times n x) (times n y))
                (quotient x y)
)
)
    ((use (q-bridge)))
)

Afer this proof has been completed, all but the last lemma should be disabled, since they are only a special case of quotient-times-times. In other proof efforts, some of the lemmas may be quite useful, and should therefore be kept. (Boyer and Moore, 1988) contains several examples.

2.5 Synthesis method

The specification to be used in the synthesis process must already have been written in the STRICT language (Campbell *et al*, 1985), a proprietary hardware description language similar to VHDL (IEEE Manual, 1987). In STRICT, the behavioural specification for each block is captured in a set of functional expressions, along with temporal information. The tool transforms this specification into a functional tree. This tree is drawn on a screen by a graphical subsystem with which the designer can subsequently interact. In order to ensure that all interactions preserve the correctness of the design, only changes that correspond to a set of formal transformations are allowed, and these must first have been verified by the Boyer Moore theorem prover.

These transformations are kept in two libraries, one generated by the user of the tool (for use with the current user defined specification), and the other one a standard library of rewrite rules which can be applied to the set of operators built in to the STRICT language. Both libraries come in the form of a text file, and their contents must be acceptable to the Boyer Moore prover (it is the user's responsibility to ensure that they are). Both libraries are therefore generated offline, before they synthesis tool can be used. The standard library contains rules relating to the STRICT operators shown in Figure 1.

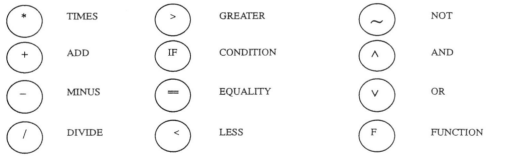

Figure 1 Standard library operations

The library is arranged in sections where each section corresponds to one of the nodes shown in Figure 1. If, for example, the designer clicks on the + node on the screen, the rewrite rules relating to the add node are made available. A typical example of such a rule might be

(equal   (plus a (plus b c))
         (plus a b c))

which simplifies the functional tree by removing one plus operation.

A behavioural specification in a high level language will usually not correspond to the most efficient hardware implementation. It will generally be necessary to modify its functional tree in order to improve the efficiency. Since only changes can be made that have first been verified by the prover, all modifications are by definition correct. The theorem prover therefore ensures that all changes made are carried out within a formal framework. Once the designer has completed his work on the functional tree, hardware can be allocated, by fetching the appropriate modules corresponding to the various parts of the design from a hardware library.

When the transformations and hardware allocation have been completed, a STRICT description of the complete design is generated, and the final layout can then be generated using standard floorplanning and routing tools.

The interactive interface of the synthesis tool is shown in Figure 2. The functional tree for one level in the design hierarchy is displayed in the centre of the screen. Below it is its lisp description. Interaction with the functional tree takes place by clicking on the icons which are situated around the edges of the screen. To access the subtrees of the functional tree, a small list of icons on the right hand side of the screen is available. It begins with eva and ends in ret (which stands for return, and which enables the user to move back up the design hierarchy). To carry out modifications to the tree the THSRCH icon at the bottom left of the screen

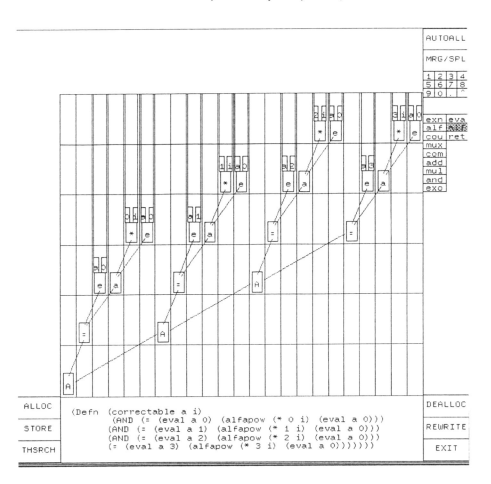

Figure 2 Interactive interface of synthesis tool

is selected, followed by a node on the functional tree. The system respoonds by searching the rewrite rule library section associated with the chosen node and selects a list of applicable rewrite rules which are presented to the user as options. If the user wishes to carry out the modification, he selects it directly and applies the change. The STORE icon which is above the THSRCH icon can be used to store a particular rule which the designer may wish to apply more than once. In this case the rule is passed to a small buffer where it can be selected and applied without searching the library for it. The REWRITE icon at the bottom right hand corner of the screen is used for rewriting a subtree within a functional tree. The ALLOC and DEALLOC icons are used for the purpose of allocating hardware to the functional tree. Allocation for a particular node on the functional tree is done by clicking on the ALLOC icon followed by the selection of a node on the functional tree. The system responds by searching a library of hardware modules each of which has an associated behavioural description. A list of modules with area and time information is provided which are guaranteed to implement the behaviour at the chosen node. The user then selects a module, and the functional tree is modified at the node where the implementation is carried out by exchanging it for a new node representing the hardware module. To deallocate, the DEALLOC icon is selected followed by a previously allocated node. In this case the node is replaced by the subtree representing the behaviour of the particular hardware module. Allocation of a functional tree can be carried out automatically by clicking on the AUTO-ALL icon. This causes the tool to search the library of hardware modules and map them directly to the functional tree. The MRG/SPL icon is used for space-time transformations. Finally the EXIT icon is used for exiting from the synthesis tool.

To apply a change to a particular point, a node from the tree is selected. The change must be in the form of a Boyer Moore rewrite rule, selected from one of the available libraries. The left hand side of this rule is matched against the tree from the chosen node (in the upward direction). If a match is found the change will then be applied. The right hand side of the rewrite rule containing the new structure is substituted in place of the old structure, and all connecting nodes are appended to the new section of tree.

Some designs may be of a regular nature and have a repetitive structure which can only be modified efficiently by applying a change repetitively throughout the design. For this purpose a feature is provided which allows changes to be made globally throughout the design.

Changes made to the original functional tree by the designer are recorded, by storing them in a text file. This is useful for a number of reasons. It enables the designer to check upon his own modifications once the design is completed. Also, whilst changes are being made to the original specification by applying rewrite rules, the theorem prover may generate new rules which the designer may wish to keep. For example, a larger rewrite rule may result from a series of smaller changes, or the designer may be able to derive a new rewrite rule. If such a rewrite rule is stored, it could easily be added to the user library at a later point.

Allocation of hardware is split into two stages, manual and automatic. The manual stage concerns the binding of operational units to the operators on the functional tree. For this purpose a library of hardware components is available, so that the designer can choose from a possibly large set Area and time information is shown in graphical form at the top of the screen as he chooses his components and allocates them.

2.6 *Synthesis example*

The screen of Figure 2 shows an example function which was taken from a error decoder module described by Kalker (Kalker, 1988). The function is called 'correctable' and would be defined in STRICT as follows:

correctable(a: byte[32], i: integer):BOOLEAN::=
(eval(a, 0) ==   alfapow(0*i, eval(a, 0)))   AND
(eval(a, 1) ==   alfapow(1*i, eval(a, 0)))   AND
(eval(a, 2) ==   alfapow(2*i, eval(a, 0)))   AND
(eval(a, 3) ==   alfapow(3*i, eval(a, 0)))

The bottom of the screen shows the S-expression equivalent of this function. We demonstrate the operation of the tool by showing how one can make this function more efficient by making formal transformations.

Any modifications that might be applied to the tree rely upon the use of rewrite rules from the library, such as:

1 / (equal ((times i 0) 0))
2 / (equal (equal a a) t))
3 / (equal ((and t a) a))
4 / (equal (and a (and b c))
         (and (and a b) c)))

In the rest of this section we will frequently refer back to these rewrite rules.

We turn our attention to the leftmost subtree on the screen of Figure 2. After clicking on the multiplication node, the screen of Figure 3 appears.

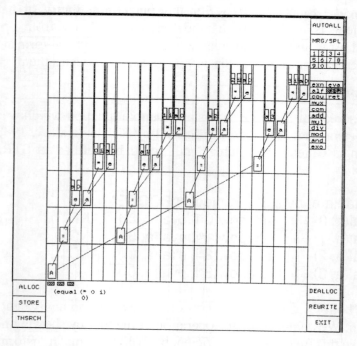

Figure 3   Error decoder example

Near the bottom of the screen 3 boxes have appeared, indicating the fact that three rewrite rules from the library are applicable at this point. The first of these rules is printed below the boxes, and by clicking on each of the boxes in turn, the designer can cycle through them. Application of rule 1 (which says that 0*i equals 0) will result in deletion of the multiplication subtree. This is shown in Figure 4. We then have a subtree representing alfapow(0,eval(a,0)), which can be replaced by eval(a,0) by rewriting. The rewrite rule is shown at the bottom of Figure 5, and the result of the application of the rule in Figure 6. The resulting subtree has two equal branches, so rule 2 can be applied (see Figure 7) to the = node to give the result T. This is shown in Figure 8. Since ANDing with T is a no-op operation (see bottom of Figure 9), rule 3 can be applied to the AND node, resulting in the removal of the leftmost subtree.

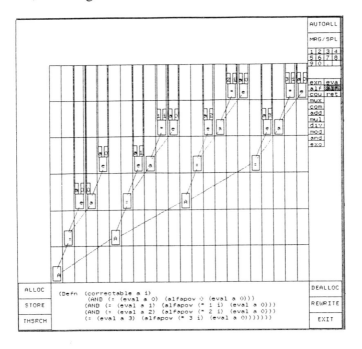

Figure 4 Error decoder example (cont.)

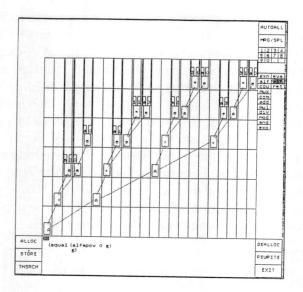

Figure 5   Error decoder example (cont.)

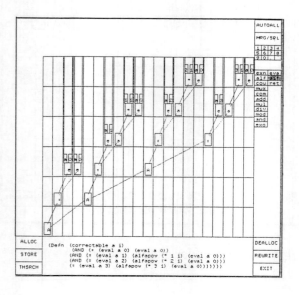

Figure 6   Error decoder example (cont.)

Use of a theorem prover for transformational synthesis 41

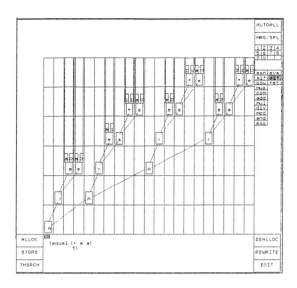

Figure 7 Error decoder example (cont.)

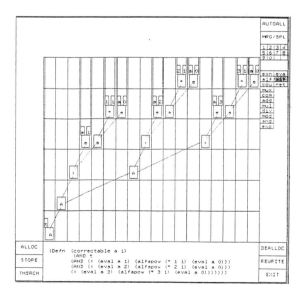

Figure 8 Error decoder example (cont.)

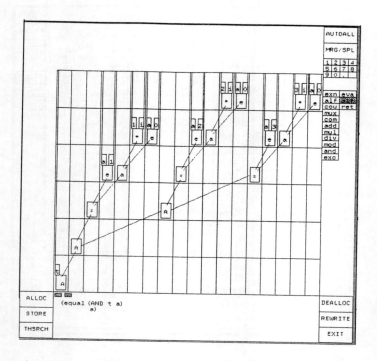

Figure 9 Error decoder example (cont.)

The final functional tree after the above changes have been applied is shown in Figure 10. The leftmost subtree has disappeared completely as a result of the transformations applied, thereby producing a more efficient design. The updated version of the 'correctable' function is displayed near the bottom of the screen.

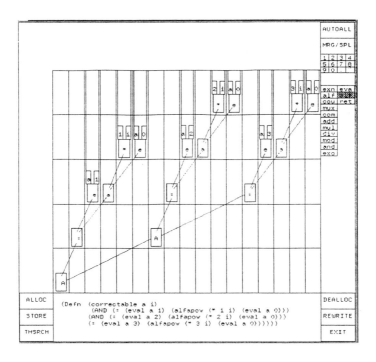

Figure 10 Error decoder example (cont.)

At this point, it will be possible for the designer to allocate actual hardware. This is shown in Figure 11, where nodes that have been allocated are shown as shaded boxes. At the top of the screen, for each box an estimate of the area and the speed for each node is shown. These in turn may prompt the designer to select a particular node, for example, one with a very large area, in order to do further optimisation. A block diagram of the resulting implementation is shown in Figure 12. We have abbreviated eval(a,0) to E0 etc., and alfapow(1*i, eval(a, 0)) to A1 etc. Upon exit, the tool will generate a complete structural STRICT description of the design, which can subsequently be used to generate its final layout.

44 Use of a theorem prover for transformational synthesis

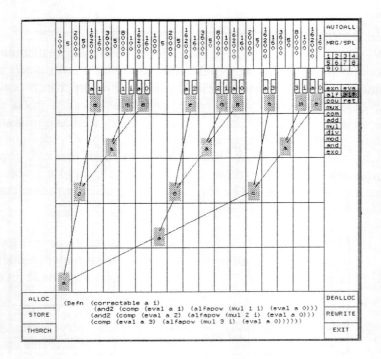

Figure 11 Error decoder example (cont.)

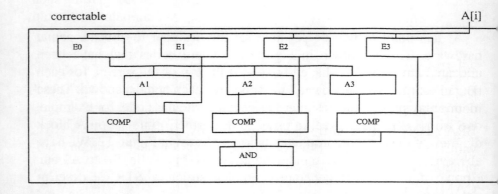

Figure 12 Block diagram of implementation

2.7 Conclusions

We have presented a prototype high level synthesis tool that integrates a theorem prover. The tool enables the designer to move towards an intended goal quickly and efficiently. The formal approach prevents a large class of human design errors and takes away a good deal of unnecessary effort on the part of the designer, as well as eliminating the need for much of the simulation now used to verify designs before fabrication. Whilst we have shown that a theorem prover can provide considerable benefits in the design of hardware, there are some areas where further work should be done to gain the maximum advantage from the use of formal methods. These include the following:

1) At present the prover has to be run separately from the tool. This situation is difficult to avoid because it runs under an interpreted Lisp environment. We are investigating how to couple them more closely.

2) The handling of the lemma libraries is very basic. A more sophisticated library interface would increase the ease with which the tool can be applied to real designs.

3) Timing aspects of a design are not described formally - possible changes are only functional, and the transformations do not take into account constraints that need to be adhered to regarding such timings. Correct combined time and space manipulations would also be useful.

2.8 References

BOYER R.S., MOORE J.S., "A Computational Logic Handbook", Academic Press, 1988.

BRAYTON R.K., CAMPOSANO R., DeMICHELI G., OTTEN R.H., vanEIJNDHOVEN, J., "The Yorktown Silicon Compiler", Silicon Compilation, Gajski (Ed), Addision Wesley, 1988.

CAMPBELL R.H., KOELMANS A.M., McLAUCHLAN M.R., "STRICT A Design Language for Strongly Typed Recursive Integrated Circuits. IEE Proceedings, Volume 132, Pts E and I, No 2, 1985.

CAMPOSANO R., "Behaviour-Preserving Transformations for High Level Synthesis", Lecture Notes in Computer Science 408, Springer Verlag, pp106-128.

DeMAN H., RABAEY J., SIX P., ClAESEN L., "Cathedral II: Silicon Compiler for Digital Signal Processing.", IEEE Design and Test, Vol 3, No 6, December, 1986

DENYER B.P., "SAGE, A Methodology and Toolset for Architectural Synthesis", Technical Report, Department of Electrical Engineering, Edinburgh University, 1989.

ElMASRY M.I., BUSET O.A., "ACE, A Hierarchical Graphical Interface for Architectural Synthesis. Proceedings 26th ACM/IEEE Design Auto. Conference, 1989.

FINN M., FOURMAN M.P., FRANCIS M., HARRIS R., "Formal System Design - Interactive Synthesis Based on Computer-Assisted Formal Reasoning", Proceedings IFIP Workshop - Applied Methods for Correct VLSI Design, Belgium, 1989.

GORDON M. "HOL, A Proof Generating System for Higher Order Logic" - VLSI Specification, Verification and Synthesis", G. Birtwhistle and P. A. Subramanyam (Eds), Kluwer, 1987.

HUNT W., "FM8501: A Verified Microprocessor", PhD Thesis, University of Texas at Austin, 1985.

IEEE Standard VHDL Language Reference Manual, IEEE Press, 1987.

KALKER T., "HOL Semantics for DSP, Philips Research Labs Technical Report, Eindhoven, The Netherlands, 1988.

McFARLAND S.J., PARKER A.C., CAMPOSANO R., "Tutorial on High Level Synthesis", Proceedings 25th ACM/.IEEE Design Auto. Conf, 1988.

VERKEST D., JOHANNES P., CLAESEN L., DeMAN H., "Formal Techniques for Proving Correctness of Parameterised Hardware Using Correctness Preserving Transformations", Proceedings IFIP Workshop - The Fusion of Hardware Design and Verification", North Holland, 1988.

Chapter 3
An Overview of High Level Synthesis Technologies for Digital ASICs

K.W.Turnbull

3.1 Introduction

To design electronic systems at the printed circuit board and chip level, a "top-down" approach is required to manage the complexity of the problem. A hierarchical approach of partitioning and refining the component descriptions at each level is used. The goal is to arrive at a description which can be realised in the form of gate arrays, programmable logic devices, custom chips or standard parts.

Synthesis has come to mean the process of automatically transforming a description at one level of abstraction into a more detailed representation at a lower level. In this way one or more synthesis steps can produce a physically realisable design from an abstract description.

The abstraction levels used in what is termed high level and logic synthesis depend on the design methodology and particular tools available, but approximate to the following stages: A system level specification of function plus timing requirements; a partitioned system description consisting of a netlist of chips with architectural or algorithmic descriptions; a chip level description refined to an intermediate register transfer level(RTL); a chip level logic gate description with timing; a chip level gate description with back annotated physical layout data based timing. The stages after this can be thought of as physical synthesis and have in general been comprehensively automated during the 1980's. They consist of cell synthesis, placement, routing and transformation to mask data steps.

The term high level synthesis is often applied to the transformation from algorithmic to RTL level. RTL synthesis is the process of optimising the RTL level description. Logic synthesis is the process of transforming optimised RTL descriptions into a logic gate realisation. This gate level can be optimised for desired speed and area using a particular target library and is therefore very appropriate for gate arrays and standard cell realisations. However, a further physical synthesis stage of cell compilation might be considered for custom chips. See Figure 1.

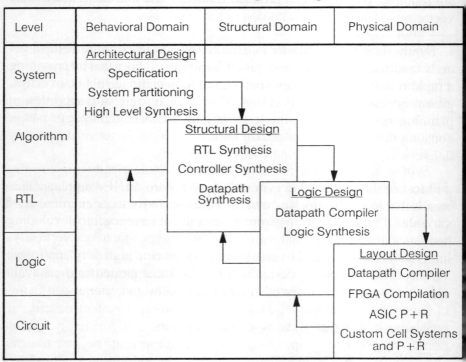

Figure 1 Design phases in a top-down language driven design methodology

Verification that synthesis transformations have preserved the intended behaviour is generally required. Interface behaviour can be verified by comparing simulations of the different levels. Structural verification of transformations can also be performed. There are developments in formal correct-by-construction systems where each transformation is mathematically defined so as not to change the overall behaviour. However, because each higher abstraction level approximates some of the information at the levels below, one has to carefully bound such approximations to preserve the validity of such systems.

The need to test chips during manufacturing means that test vector sets are required. Synthesis can produce testable logic and the required test vector sets. This process is known as test synthesis.

Synthesis helps a designer to explore the design possibilities and get early feedback on the implementation feasibility. It does this by providing a rapid route to a technology-specific implementation. This is, in effect, giving access to the "bottom-up" view of a design. Advance design planning of architectural implementation, optimal technology, power consumption, area and final packaging are important factors in decreasing delivery times and producing cost-effective parts.

Most of the problems in synthesis are NP-hard or NP-complete. This means that exact solutions cannot be produced above a certain number of variables. This has led to an enormous amount of research into developing heuristics. It also means that many solutions are specific to one vertically-integrated design style. The challenge is to develop generally applicable theories, methods and tools, which can be incorporated into practical application specific integrated circuit (ASIC) design systems.

Synthesis systems can be be categorised into:
- System synthesis
- Interface synthesis
- High level synthesis
- Formal synthesis
- RTL synthesis
- Logic synthesis
- Test synthesis
- Physical synthesis

The discussion of synthesis techniques will be limited to high level methods. However, this will be preceded by a summary of the benefits of synthesis.

3.2 The Benefits of Synthesis

The reasons for using synthesis systems are:

1) The number of possible implementations of a chip's design represents a solution space which is time-consuming to explore without automated Computer Aided Design (CAD) tools. In the absence of a synthesis system, a designer will tend to work in a certain niche style. This can lead to conservative solutions which do not use new technologies, components or architectures.

2) For a given design style they reduce the design cycle by automating many of the design procedures.

3) For a given design cycle they enable more time to be spent on investigating alternative design implementations, from architecture down to technology. This enables more cost versus feature trade-off analysis to be done.

4) By encouraging a higher level design style, design intent can be separated from design implementation.

5) They enable an existing design to be quickly re-implemented in a cheaper or faster way. (A critical factor in extending the market life of established products). Currently this is a dominant requirement [TRG, 1991].

6) The system becomes a self-documenting repository of design techniques. Expertise can be added into the system via sets of tools which encapsulate design specific heuristics and algorithms.

7) The use of formal representations throughout the design process allows verification procedures to be run, resulting in fewer errors and a constant reference to design intent.

8) Cheap, programmable silicon structures are now available for medium-sized designs. By targeting synthesis systems at such devices new designs can be produced cheaply and quickly.

9) By generating testable structure and vectors as part of the design process, time consuming chip level test generation and fault simulation is no longer essential. Highly testable chips with guaranteed fault coverage can also be produced.

3.3 System Level Synthesis

The design of a system starts as a procedural specification plus a set of interface constraints. System level synthesis is the partitioning of systems into realisable sub-systems (McFarland, 1990). More fully, it is the process of splitting a system into sets of processes, and specifying the interfaces between them. At this level, tradeoffs are made using existing high level components and synthesised ASIC solutions. Few systems support this. Currently, an expert system (MICON) has been used for a board level approach, based only on a library of high level standard components. A system partitioner, APARTY, has been used in SAW(Thomas et al, 1990), and similarly CHOP in USC (Parker (b) et al, 1991). Many systems reduce the complexity of the problem and simply solve one style of system implementation in a vertically integrated manner, for example, CATHEDRAL for digital signal processing (DSP). More flexible synthesis systems are gradually emerging.

3.4 Interface Synthesis

Interface synthesis as outlined above has become a separate area of investigation. The problem is to interface mixed synchronous and asynchronous processes. Essentially it is a control flow problem. Timing constraints are critical. This is in contrast with high level designs which are more cycle-based with weak timing constraints. Poor timing support in VHDL has lead to a rash of languages in interface work, for example,

SLIDE (Hayati *et al*, 1988), VAL (Augustin *et al*, 1991) ISYN (extended ISPS (Thomas*et al*, 1990)). With the development of asynchronous state machine research and the use of internal on chip tristate buses, this issue is now becoming important for internal chip synthesis. See (Borriello, 1991) for more detail.

3.5 High Level Synthesis

High level synthesis is the step of taking each of the partitioned set of algorithmic implementations and producing a target RTL description in which the sequencing and type of data flow operations have been fully specified. This target netlist consists of a number of buses or random multiplexed interconnections, storage in the form of registers, RAM or ROM, a clocking scheme, functional units, and control logic.

An overview of the process is shown in Figure 2.

Steps that exist in a high level synthesis system are:

Translation:

- Design represented in suitable input language
- Translation to intermediate representation

Optimisation:

- Compiler optimisations
- High level floorplanning
- Pipeline optimisation analysis
- Operation scheduling of functional units
- Functional unit allocation
- Datapath allocation

3.5.1 Translation to intermediate representation

There are two general targets; parse tree or data flow graph.

An overview of high level synthesis technologies 53

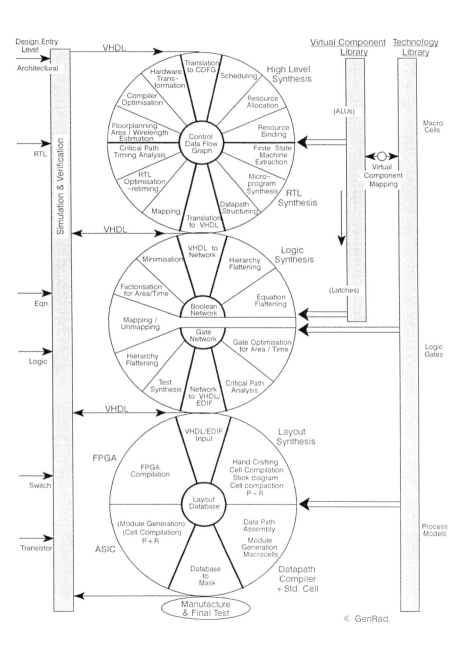

Figure 2 Synthesis operating overview

3.5.1.1 Parse tree representation

Although most simple software compilers operate on tree data, this is not generally used in synthesis since most of the sub-problems end up being formulated as graph problems. Therefore, the initial parse tree is generally converted to a flow graph after performing simple in-line optimisation.

3.5.1.2 Control Data Flow Graph (CDFG)

The aim of this representation is to identify the data and control flow relationships. The general definition is G=(V,E) where V is a set of nodes and E is a set of directed edges. The nodes of the data flow graph are operators, while edges represent data. A node takes data, generated by predecessor nodes from its incoming edges, and generates new data for successor nodes, onto its outgoing edges. Therefore the order of the nodes gives an ordering of operations to be performed on the time data flowing through the graph network. At no time can an edge carry more than one data item. This property allows a correct modelling of hardware operation (Jong, 1991, Stok, 1991 and GenRadSIF). (Note that modelling of a CDFG in VHDL has been done (Madsen and Barge, 1990), thus enabling active simulation and debugging using conventional VHDL tools.)

Basic CFDG node types are:

1) *Operator nodes*. These can be basic boolean operators or more complex data flow operators, which, in themselves, can be sub data flow graphs.

2) *Input, output or data nodes* for read and write operations.

3) *Control nodes*. These are branch(fork) and merge (join) nodes. Branch and merge nodes represent if/then/else, loop/endloop and case (switch) statements. These have two incoming edges, a data and a control edge plus two or more outgoing data edges. The incoming edge selection is determined by the control edge. Merge nodes have one outgoing and incoming control edge and

several incoming data edges. Similarly, the path is determined by the control edge. In a loop there is a data edge to the test operator (the value of the controlling loop variable) from the merge (for the branch in the case of if/then/else). This enables the test operator to monitor the loop correctly. The control edge for the merge and branch nodes comes from the same test node.

4) *Constant input nodes* which hold time invariant data.

There can be many variations in the graph representations. Two other categories that have been discussed (Stok, 1991) are semi-data flow and separated data and control flow graphs. A semi-data flow graph is defined as G(V,W,E) where G is a bipartite graph, node set V represents operations and node set W the variables. The edges, E, represent a mapping from variables to operator inputs and operator outputs to variables.

In the separated case, the control graph is defined as CFG=(Vc,Ec), where vertices Vc represent operations and edges Ec represent a simple precedence. The data flow graph is defined as DFG =({Vc,Vd},Ed), where vertices Vd represent variables and edges Ed represent data dependencies. Data precedence is simply that output data always follows input data. An example of the control and data flow graphs associated with a fragment of VHDL is shown in Figure 3.

Formal software specification systems have followed this combination of control and data flow. For example, IP nodes(Schindler, 1990) are now being used for CASE software specification. In the past, non machine-analysable specification systems used either data flow or control flow graphs, but not both.

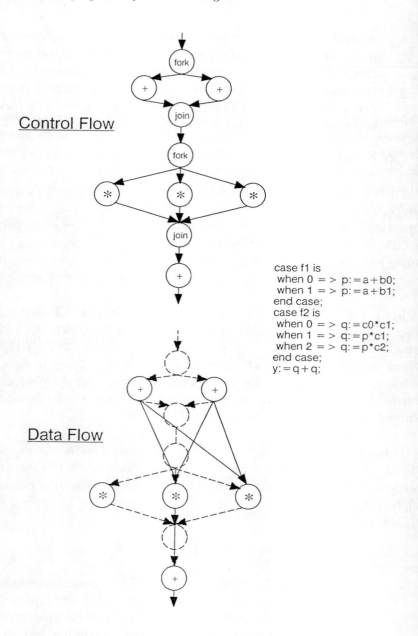

Figure 3 Control flow and data flow example

3.5.2 Optimisation

3.5.2.1 Compiler optimisations

Many classical optimisations as developed for software compilers can be used. Optimisations which always reduce code complexity are performed straightaway. However some optimisations which directly affect the hardware configuration are incorporated into the later scheduling or allocation steps. The following optimisations are used:

1) In-line expansion of procedures. This is done immediately unless the designer wants to introduce a design partition.

2) Constant folding

   e.g. a+3+2 = A +5.

   This is straightforward for a block. However global folding (i.e across loop boundaries) requires data flow analysis.

   e.g. a:=1;
   loop
   a:=a+1;
   endloop;

   here it is not correct to replace a:=a+1; with a:=2;

3) Constant propagation

   e.g. pi=3.14; a=pi/3; => pi=3.14; a=3.14/3;

4) Common subexpression elimination. This re-uses intermediate results which are used repeatedly without being updated. It requires flowgraph analysis to supply variable lifetime information. Subexpression elimination affects scheduling and allocation(see later) and so is likely to be deferred.

5) Frequency reduction, where fixed assignments are taken outside loops

6) Dead code removal where unused code is discarded

7) Loop unrolling

8) Transformation of operators (strength reduction). This is where operators are replaced by those which are more efficient in space or time

e.g.
X**2 => X * X,
J*2 => shift_left(J,1).

Transformation choices are dependent on the target hardware. In high level synthesis an extra degree of freedom over software optimisation is the ability to generate the target hardware rather than using a fixed architecture. In the example there is the potential area saving of having a bit shifter rather than a multiplier. Clearly in the specific case it depends whether such a potentially expensive resource as the multiplier has to be implemented for other reasons and in that case, whether it has some spare cycles available for the required operation.

9) Minimum bit representation. Here integer sub ranges are translated to minimum bit patterns. Otherwise a hardware implementation can end up performing 32 bit operations if default integer widths are used. This is not such a problem in software, where operations are normally more efficient when targeted to the native word length of the target machine.

Global flowgraph analysis. Optimising across code block borders is a classically hard, non-local optimising problem (Tremblay and Sorenson, 1989). Flowgraph analysis of the CDFG enables calculation of variable lifetimes, reaching definitions, available expressions and very busy expressions (Tremblay and Sorenson, 1989). Variable lifetime information is required for scheduling and allocation. In global flow analysis, looping expressions have to be dealt with. The loops can be cut by using definitions of strongly connected regions (Tremblay and Sorenson, 1989). Region based optimisation offers O(nlog(n)) performance. Iterative methods are O(n**2). (Ullman, 1973) offers algorithms for O(n) performance.

3.5.2.2 Transformation

This is the process of splitting or collapsing nodes of the CDFG. Some of these operations occur as part of the compiler frontend e.g., loop unrolling, in line code expansion. Splitting paths produces concurrent processes. Pipelining is a specialised case of concurrent process creation.

3.5.2.3 Scheduling

Scheduling is the process of determining which operation will occur in which control step. Conventionally, a control step is a clock cycle for a synchronous machine. Within a control step several fast operations may take place. This is known as chaining. The starting point for most scheduling algorithms is to run fast, greedy, constructive algorithms which bound the problem. As soon as possible (ASAP) scheduling calculates the earliest possible control step in which an operation can occur. As late as possible (ALAP) gives the latest time an operation can occur. The two limits determine the mobility of each operation.

The problem, then, is to minimise the hardware requirement for a given number of time steps (time-constrained scheduling), or to find the fastest schedule for a given amount of hardware (resource constrained scheduling).

A further complication arises when there is a choice of operators each with their own sets of intrinsic per-cycle delays, physical area, and number of execution cycles. This requires some consideration of functional unit allocation before or together with scheduling. For example, serial operators require extra control steps. This results in additional multiplexors and, possibly, additional registers for temporary variable storage (Fuhrman, 1991). Operator types can be pre-determined to minimise complexity. The alternatives can then be individually scheduled for comparison.

Some scheduling systems predict the effect of interconnect delays by using floorplanning feedback. Interconnect delays are now significant with current sub micron technologies. The 3D scheduling part of the USC system (Parker (b) *et al*, 1991) introduces duplicate operators to preserve a critical path, rather than introduce additional interconnect. The overall active area and critical path can actually be improved by adding duplicate operators. Here the allocation and binding are also incorporated to enable better estimation of the wiring overhead.

Scheduling can also be carried out using branch-and-bound (Berry and Prangrle, 1990), and mixed integer linear programming (MILP) (HwangC *et al*, 1991 and Hafer, 1991), but these methods are not generally suitable for large problems. Force directed scheduling has been widely used. This uses an overall cost function for each control step based on the probability of operations existing there (Paulin and Knight, 1989). Other techniques include path analysis, such as HIS (Camposano *et al*, 1991), and min-cut (Park). MAHA (Parker, 1986) uses urgency scheduling which schedules critical path operations first. BUD (McFarland, 1986) chooses as its selection criterion the length of the path from an operation to the end. Note that the scheduling and allocation problem is similar to task scheduling in project management. A treatment showing a solution using Hu's list scheduling algorithm is given in (Gould, 1988).

3.5.2.4 Pipelining

Pipelining can either be pre-specified or incorporated via operators which have latency. Normally, pipelining is specified at a system partitioning level, because of the complexities of calculating the extra control implications.

3.5.2.5 Allocation

Allocation is the process of assigning functional units. There are circular dependencies between scheduling and allocation. Ideally, scheduling and allocating should be done together. Knowing the cost of hardware (e.g., speed and area), a good scheduler can help allocation. Connectivity, multiplexor, functional unit, and register costs should be known at this stage. Allocation is made up of the following:

Operator Assignment or Binding. This is where operations, variables and data transfers are mapped to actual components. The clocking and latch scheme should be known at this stage. Sometimes this is treated as part of scheduling.

Register assignment. The problem is to minimise the number of registers required for variable storage. This can be done using data flow analysis and classical lifetime analysis. An incompatibility graph can then be created for co-existent variables. A graph colouring algorithm (Turnbull, 1989), as used as in CADDY (Campasano, 1991), or clique partitioning methods (Tseng and Siewiorek, 1983) is then used to produce disjoint sets.

Module assignment. There are generally very simple mappings of functional units, which are implicitly assigned to physical units by operator allocation. Changing the style of implementation at this stage (for example, parallel to serial) would invalidate the scheduling and allocation. Therefore this is normally fixed before scheduling and allocation.

Data path assignment/binding. Bus or data path partitioning can reduce the amount of steering logic (by grouping non-interfering sets of compatible signals). There are additional hardware implications in the choice of multiplexed or tri-state bus. Bus splitting can allow parallel operations. Implementing two short buses instead of one may have little overhead on area, but bring significant gains in performance (Pangrle *et al*, 1991).

3.5.3 Using estimation to meet constraints

High level technology mapping. Advance technology mapping at the high level can be used to increase estimation accuracy (Dutt, 1991). Essentially, this pre-compiles complex cells such as ALUs from a generic language (LEGEND) into a given technology, so that characteristics such as area and speed are known at allocation, scheduling and binding. Here a symbolic functional approach is taken; matching is done by classification of function rather than of behaviour. Another approach is to use virtual components HidesignA. This uses generic cells which can can be mapped to a particular library. VHDL also allows the use of packages to specify operator overloading for immediate operator macro expansion or mapping to virtual components. A further method in VHDL is to use parameterised functions.

Floorplanning. This is the trial placement of given or estimated modules to check that physical limitations (total chip area, wire delays, etc.) are not violated. It can also be used at the system synthesis level by inclusion of a system partitioning to produce estimated block data. Partitioning at the system level is more difficult than conventional structural netlist because different architectures may have radically different interconnectivity and module sizes. The USC system is a good example of using floorplanning [Parker (a) and (b) *et a*l, 1991).

Design constraints. Constraints, or costs, of synthesis operations include total area (modules + wiring), worst case cycle time (module + wiring delay) and total number of execution cycles (operation cycles and sequencing). Although high level synthesis is sometimes said to be a technology independent level, good results do not usually occur unless reasonable technology-specific estimates of area and speed are fed into the process. Scheduling cannot be usefully performed without some knowledge of the cycle times and delays of target units. Floorplanning is also useful for advance wire estimation as discussed in (Parker (a) and (b) *et al*, 1991).

3.5.4 Formal methods

Formal systems were originally developed for specification and verification. They allow hierarchical systems of rules or transformations to be built up. One of the first practical applications was the prolog program VERIFY (Barrow, 1984). This was actually used to verify the function of a 30k gate image-processing chip within Fairchild. Systems such as HOL (Gordon, 1988) have spawned a whole branch of research. These can be used as design verification or synthesis systems (Mayger and Fourman, 1991) if a rule base or manual guidance is supplied to limit the design space explosion as the design is expanded out to primitives. In the past, these systems have been weak on scheduling, although the application of MILP methods (see below) is now addressing this. Industrial systems are beginning to use these methods (Kalker, 1991 and Mayger and Fourman, 1991). There are now binary decision diagram (BDD) based verification systems such as PARIS/PRIAM (Madre, 1990) and more refined hierarchical systems (Burch et al, 1991).

3.5.5 Timing constraints and analysis.

VHDL timing is being extended in the VHDL '92 proposals (VHDLLDR). However, many systems for supporting ASIC delays in VHDL have been proposed (Yeung and Rees).

Given the ability to input timing constraints, (Zahir and Fichtner, 1991) gives an excellent discussion and proposed solution for analysing detailed timing at the high level before translating to a low level netlist. A modified bipartite CDFG that contains signal nodes and operation nodes is used. A timing graph for functional units is used which has nodes representing events and edges representing timing constraints. A forward edge with positive weight gives the minimum constraint and a backward edge with negative weight gives the maximum constraint. Solving this constraint path is the critical path problem of assigning earliest and latest event times (and hence calculating slack) in $O(|Vertices|.|Edges|)$ time.

3.5.6 Algorithmic techniques

A quick review of the main techniques used in high level synthesis follows:

3.5.6.1 Hierarchical clustering

Clustering is similar to clique partitioning in that it is a process of merging strongly connected nodes. It is a fuzzy technique, not an explicit relationship. Given a graph with edges representing some relationship between the nodes, the clustering ratio between nodes is, for a node i, and one of its neighbours, j:

$F(Connectivity\ between\ nodes\ i\ and\ j)$
$G(Total\ External\ Connectivity\ of\ i)$

Nodes with the highest ratio (e.g., a node with only 1 edge connection) are clustered to the relevant neighbour and are considered as one node in the next analysis iteration.

Hierachical or multi-stage clustering is a variation that allows different clustering criteria to be applied at each clustering stage. The criteria used are control flow and data flow between subgraphs, schedule lengths and operator similarities. Clustering is a bottom-up process as opposed to top-down methods such as min-cut. Multi-stage clustering is used for chip partitioning in APARTY (Thomas *et al*, 1990).

3.5.6.2 Min-cut partitioning

Min-cut partitioning methods are top-down. They generally divide the nodes of the graph into equally sized partitions (usually 2 or 4 at a time). The heuristic algorithms are based on (Kernighan and Lin, 1970 and Fidducia and Matheyses, 1982). A weakness is that a greedy (as opposed to global) strategy is used. Thus one may not get the overall best result by using initial partitions with the least edge communication. Moreover, only connections within a partition are considered. (Dunlop and Kernighan, 1985) extended the method with "terminal propagation", which takes account of the existing external partitions. A good overview of partitioning and clustering is given in (Sangiovanni-Vincentelli, 1987).

3.5.6.3 Clique partitioning

This is a standard graph technique covered in (Gould, 1988) and used in systems such as FACET (Tseng and Siewiorek, 1983). It is the process of dividing the nodes of a graph into a minimum number of disjoint clusters. Each node appears in only one cluster and each cluster is a complete graph. This technique is used when finding operations that nodes can share without conflict. Thus it is suitable for allocation. The graph is set up by having edges between nodes (operations, data or interconnect) that can share the same hardware.

3.5.6.4 Graph colouring

Graph colouring attempts to use the minimum number of colours when assigning a colour to a node of a graph so that nodes of the same colour have no connecting edges. This is a complementary operation to clique partitioning where the minimum number of clusters containing nodes which have no common edges is sought. This is used for representing exclusive behaviour of nodes (i.e., nodes that can execute in parallel). Again see (Gould, 1988) for a general treatment and CADDY (Campasano *et al*, 1991) for a specific usage.

3.5.6.5 Mixed integer linear programming (MILP)

This attempts to optimise a set of global constraint equations and results in a resource constrained scheduling problem (Hafer, 1991), (HwangC *et al*, 1991). The computational costs are high, although breakthroughs are claimed with OASIC (Gebotys and Elmasry, 1991).

3.5.6.6 Condition vector node encoding

A global analysis of the CDFG, as in HIS (Campasano, 1991), can suffer from an exponential increase in the number of paths, especially those with nested control constructs. A speed-up for dealing with conditional branches is described in the CYBER system (Campasano, 1991). This uses condition vectors attached to all nodes and is a means of working out mutual exclusiveness which is not apparent at the behavioural level. Simply, the condition vector is an encoding at each node such that if

66 An overview of high level synthesis technologies

$$CV_i \wedge CV_j = 0$$

then the operations cannot co-exist at the same time and can therefore share resources. The coding is organised to allow very rapid pruning of the graph paths. The VHDL example in Figures 3 and 4 shows two sets of mutually exclusive operations which can share resources. Only one "+" and one "*" operator is required. Note, no vector encoding is shown in the diagrams.

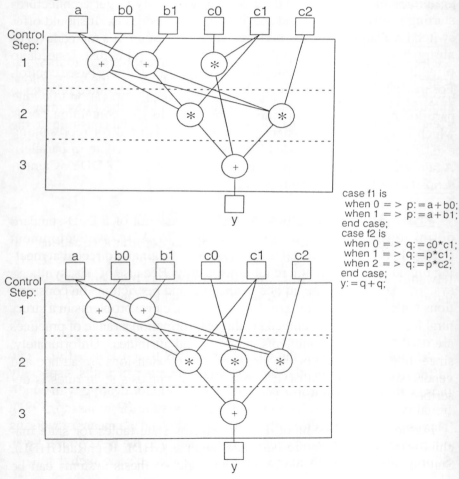

```
case f1 is
  when 0 => p:=a+b0;
  when 1 => p:=a+b1;
end case;
case f2 is
  when 0 => q:=c0*c1;
  when 1 => q:=p*c1;
  when 2 => q:=p*c2;
end case;
y:=q+q;
```

Figure 4 ASAP and ALAP operation scheduling

3.6 Synthesis Specification Language Considerations

3.6.1 General requirements

The functions required from a high level language are support for simulation, synthesis and formal verification. It should have standardisation controls for common usage. There are software specific requirements, such as no GOTOs for flow graph analysis, or non ambiguous grammars for parsing. It should be easy to specify target architectures starting from procedural and specific timing constraints. It should offer switch level and analogue interfaces. It needs to support a mixed level of abstraction, from behavioural to gate level structural netlist. It needs to offer the choice of procedurally based and non procedurally based control. For instance, a block of statements only needs to be executed relative to another block of statements, but within the block the order may be irrelevant and to impose an order would add an artificial constraint to the design description.

3.6.2 Using VHDL for synthesis

VHDL 87 is not ideal for synthesis. It grew out of a DoD standard definition project, in a similar fashion to ADA's development (VHDLLRM, 1987 and VHDL, 1987). It was originally directed at meeting mixed level simulation requirements only. Because of this, synthesis systems have had to use ad hoc extensions and restrictions on constructions which, while meaningful for simulation, cause problems in a structural interpretation. To embed synthesis information a range of practices are used: embedded comments, pragmas and attributes. Unfortunately, since they are not part of the language, these extensions are ad hoc and vendor-specific. A list of 12 basic VHDL synthesis style rules is described in (Camposano, 1991).

Furthermore, one cannot directly express state tables for state machines unlike more mature languages such as GHDL (GenRadGHDL). State machines extractable by current logic synthesis systems can be described in a variety of ways (Forrest and Edwards, 1990). A preferred form for synthesis systems is where the flow is controlled by state transitions generated in a separate process. The combinatorial and sequential parts are thus separated.

For example:

```
ARCHITECTURE behaviour OF simple IS
   TYPE state_type IS (state0,state1);
   SIGNAL current_state,next_state: state_type;

-- Sequential Clock part
-- Clocking scheme is implicitly
-- rising edge triggered
BEGIN
   state_transition:PROCESS
BEGIN
   WAIT UNTIL clock'EVENT AND clock='1';
   current_state <= next_state;
END PROCESS;

-- Combinatorial part
-- sets output out1 and next state depending on
-- current state and input in1
transformation_functions: PROCESS (current_state)
CASE current_state IS
     WHEN state0 => out1<='0';
        IF in1='0' THEN
             next_state  <= state0;
                ELSE
             next_state <= state1;
        END IF;
     WHEN state1 = > out1<= '1';
        IF in1='0' THEN
             next_state <= state1;
```

```
            ELSE
               next_state <= state0;
            END IF;
        END CASE;
    END PROCESS;
END BEHAVIOUR;
```

This is an artificial restriction on the expression of state machines. Ideally the state machine should be extracted from global analysis of such a description. There are systems that already allow specification of the control flow within the data path (Harper *et al*, 1989). VHDL 87 does not explicitly support clocking schemes for state machines unless a gate level implementation is used. Timing is also not explicitly supported in a form suitable for synthesis. VHDL 87 only supports event-cause constructs. This can over-constrain synthesis systems. One also needs relativity and duration support (Dutt, 1990). A solution for current VHDL 87 has been proposed by (Yeung, 1991) amongst others.

VHDL 87 cannot support accurate library modelling, unlike existing languages such as GHD (GenRad), which can combine accurate pin-pin delays, user definable delay functions, truth tables and state tables.

The VHDL standards group, VASG, is addressing many of these points in VHDL '92. More explicit and standardised attributes, a more sophisticated time model and back annotation will be supported.

3.6.3 Other synthesis languages

3.6.3.1 Research

Many research groups have developed specialised languages (mostly pre-dating VHDL). Examples are ISPS from CMU, DDL(Duley, 1968), BIF (Dutt, 1990), Spec charts (Narayan et al, 1989), HardwareC (DeMichelli,HLS), Silage from Cathedral, V from IBM, Logic-III from (OASYS) and HSL-FX from NTT. Some languages explore particular formalisms and semantics such as HOL (Gordon, 1988) for formal verification, others have grown out of other projects, for example ISPS.

One example of a recent language is BIF (Dutt, 1990) which is used to capture design specification and then translated into VHDL. The main requirement is to allow partial structural design specification, user bindings of operators and variables to components, and to allow a mechanism for mixed manual/automatic interaction. It supports the detailed timing relationships required for interface specification timeouts, hierarchy and concurrency. The general form is of a state table but it explicitly provides for each current state: sets of conditionals; sets of actions upon entering the current state via the conditionals; the next states via the conditionals; and the next state triggering events for the conditional.

3.6.3.2 Commercial

UDL/1 is based on HSL-FX used by a consortium of Japanese companies. It was set up as a 5-year research project in 1989 to compete directly with VHDL (UDL1, 1990). Although synthesis support was one of the initial goals and is built into the first draft of the language, it does not receive much support outside Japan.

The proliferation of hardware specification languages is not appropriate from a user point of view and is uneconomical to support from the vendor side. So although there are deficiencies, VHDL support is demanded in the commercial arena and will remain the design interchange standard for a least the next 3-5 years, despite the exploration of synthesis intermediate formats.

3.7 *Summary*

A number of techniques for dealing with high level synthesis tasks have been covered. Unfortunately, the interaction between many of these tasks means that they need to be integrated vertically and solved in a limited design-specific way. Alternatively the tasks can be organised functionally as described above and run in an iterative fashion. This leads to the toolbox concept shown in Figure 2. Tool operations can be sequenced depending on the type of design. Sequencing can be performed manually or by an expert system.

3.8 Acknowledgments

Randeep Soin for his encouragement. Sara Turnbull for her patience. GenRad DAP and Martin Baynes for their support. All the HidesignA team, especially Steve Lim and Ping Yeung, for their discussions and expert help. James Ball for his artistic assistance. The terms "GHDL", "HidesignA", "HiTest", "Master ToolBox" and "Virtual Component" are trademarks of GenRad, Design Automation Products, UK.

3.9 References

AUGUSTIN L., LUCKHAM D. *et al* "Hardware Design and Simulation in VAL/VHDL", Kluwer Academic Publishers 1991.

BARROW H. G., "VERIFY: A Program for Providing Correctness of Digital Hardware Designs", Artificial Intelligence Journal Vol 24 pp 437-491, 1984

BERRY N. and PANGRLE B., "Schalloc: An Algorithm for Simultaneously Scheduling & Connectivity binding in a Datapath Synthesis System", Proc EDAC 1990.

BORRIELLO G., Tutorial on Interface Logic synthesis EDAC 91.

BURCH J.,CLARKE E. and LONG D., "Symbolic Model Checking with Partitioned Transition Relations", Proc VLSI'91.

CAMPOSANO R., WOLFE W., (Eds) "High level VLSI Synthesis", Kluwer Academic Publishers 1991.

DULEY J. and DIETMEYER D., " A Digital System Design Language (DDL)", IEEE trans on Comp C-17 Sept 1968.

DUNLOP A. and KERNIGHAN B. ,"A procedure for the placement of standard cell", VLSI circuits IEEE Trans on CAD vol4 no 1 Jan 1985.

DUTT, N., J. KIPPS, Bridging High level synthesis to RTL technology libraries ,Proc 28rd DAC acm/ieee June 1991.

FIDDUCIA C. and MATTHEYSES R., "A Linear Time heuristic for improving Network Partitions", Proc 19th DAC acm/ieee 1982

FORREST J., EDWARDS M. "The Representation of State Machines in VHDL", Proc 1st Euro Conf VHDL Sept 1990.

FUHRMAN T. ,"Industrial extensions to University high level synthesis tools: Making it work in the real world", Proc 28rd DAC acm/ieee June 1991.

GEBOTYS C. and ELMASRY M., "Simultaneous scheduling and allocation for cost constrained optimal architectural synthesis" Proc 28rd DAC acm/ieee June 1991.

GenRadGHDL, "GHDL Reference Manual", GenRad Design Automation Products, Fareham UK.

GenRadSIF, "SIF: Synthesis intermediate Format" GenRad DAP Internal specification.

GORDON M., "HOL: A proof generating system for higher-order logic" in VLSI Specification, Verification and Synthesis, Kluwer Academic Publishers 1988.

GOULD R., "Graph Theory" Benjamin/Cummings Publishing Company 1988.

HAFER L., "Constraint improvement for MILP-Based Hardware Synthesis", Proc 28rd DAC acm/ieee June 1991

HARPER P., KROLIKOSKI S. and LEVIa O.,"VHDL as a Synthesis Language", Proc of the 9th IFIP symposium on Computer Hardware Languages and their applications 1989.

HAYATI S., PARKER A. *et al* "Representation of Control and Timing Behaviour with Applications to Interface Synthesis", Proc ICCD 1988

HWANG C-T, *et al*, "Optimum and Heuristic Data Path Scheduling Under Resource Constraints", Proc 27th DAC, June 1991.

de JONG G. "Data Flow graphs: system specification with the most unrestricted semantics", Proc 2nd EDAC IEEE Feb 1991

KALKER T., "Formal Methods for Silicon compilation" , Proc 2nd EDAC IEEE Feb 1991

KERNIGHAN B., LIN S., "An efficient heuristic procedure for partitioning Graphs", Bell system Tech journal Jan 1970

McFARLAND M., "Using Bottom Up Design Techniques in the Synthesis of Digital hardware from abstract behavioural descriptions", Proc 23rd DAC acm/ieee June 1986.

MADRE J. C. et al, "The Formal Verification Chain at Bull", Proc. Euro ASIC 1990

MADSEN J. and BRAGE J., "Flow graph modelling using VHDL bus resolution functions", Proc 1st Euro Conf VHDL, Sept 1990.

MAYGER E. and FOURMAN M., "Integration of Formal Methods with System Design", Prov VLSI'91.

NARAYAN S., Vahid F. and Gajski D., "Translating System specifications to VHDL", Proc 2nd EDAC IEEE Feb 1991.

PANGRLE B.,BREWER F.,LOBO D. and SEAWRIGHT A., "Relevant issues in High-Level Connectivity synthesis" Proc 28rd DAC acm/ieee June 1991.

PARK I-C. and KYUNG C-M., "Fast and Near Optimal Scheduling in Automatic Datapath Synthesis", Proc 28th DAC acm/ieee june 1991.

PARKER A., PIZARRO J. and MILNAR M., "MAHA: A program for datapath synthesis", Proc 23rd DAC acm/ieee June 1986.

PARKER (a) A., GUPTA P. and HUSSAIN A., "The effects of Physical Design characteristics on the Area-Performance Tradeoff Curve", Proc 28rd DAC acm/ieee June 1991.

PARKER (b) A. *et al*, "Unified System Construction (USC)" in High level VLSI Synthesis, Kluwer Academic Publishers 1991.

PAULIN P., KNIGHT J., Scheduling and Binding algorithms for High Level Synthesis, Proc 26th DAC acm/ieee June 1989.

SANGIOVANNI-VINCENTELLI A., " Automatic Layout of Integrated Circuits", in Design Systems for VLSI circuits, Martinus Nijhoff 1987.

SCHINDLER M., "Computer-aided Software Design" John Wiley 1990.

STOK L., Tutorial on high level synthesis, EDAC 91.

TRG, Logic Synthesis report, TRG group, 1991

THOMAS D., LAGNESE E., WALKER R., NESTOR J., RAJAN J., BLACKBURN R., "Algorithmic and register level synthesis: The System Architects Workbench", Kluwer Academic Publishers 1990

TREMBLAY J-P. and SORENSON P.G., "The Theory and practice of compiler writing" , Mcgraw-Hill 1989.

TSENG C-J. and SIEWIOREK D., "FACET: A procedure for the Automated Synthesis of Digital Systems", Proc 20th DAC 1983.

TURNBULL K., VEVERIS A., JOLLY C., HARDING E.J. "LEG: Layout from Equation Generator" Internal paper National Semiconductor 1989.

ULLMAN J.D., "Fast Algorithms for the Elimination of Common subexpressions" Acta Informatica, vol 2 no 3, 1973.

VHDL87, CLSI VHDl Tutorial.

VHDLLDR, VASG, VHDL 1992 Proposals.

VHDLLRM, IEEE,VHDL Language Reference Manual 1987.

YEUNG P. and REES D., "Resource restricted global scheduling". Proc VLSI '91.

YEUNG P. and REES D., "Timing Constraint Specification in VHDL", Private Communication from University of Edinburgh.

ZAHIR R. and FICHTNER W., "Specification of timing constraints for controller synthesis" Proc 2nd EDAC IEEE Feb 1991.

Chapter 4
Simulated Annealing Based Synthesis of Fast Discrete Cosine Transform Blocks

J.P. Neil and P.B. Denyer

4.1 *Introduction*

This Chapter describes CAD techniques capable of synthesising Fast Discrete Cosine Transform (FDCT) Blocks from *behavioural*, or *algorithmic*, specifications. We introduce SAVAGE (a Simulated Annealing based VLSI Architecture GEnerator), a software tool developed under the auspices of the Silicon Architectures Research Initiative (SARI(Grant, 1990)) hosted at the University of Edinburgh.

SAVAGE is capable of taking a data-flow description of an input algorithm, and applying a number of synthesis steps, or transformations, to produce a hardware netlist of a datapath. The netlist description is then passed to logic synthesis and layout tools to complete the route to silicon. These application specific synthesis steps are controlled by the computational technique known as *simulated annealing*.

This Chapter reviews the design process, from the initial high-level description of the FDCT, through the various synthesis transformations, and presents a set of test results illustrating the flexibility of the SAVAGE software. Finally, some extensions to the prototype SAVAGE system are described.

4.2 *Problem Domain*

The large amount of information contained within a high definition digital image poses significant problems, both in terms of memory requirement and transmission latency in applications where real time, or near real time, image transmission is required.

As a result, many data compression techniques have been proposed (Chen and Smith, 1977, Wintz, 1972 and Soame, 1982). The Discrete Cosine Transform (DCT) operates on a series of blocks decomposed from the original image. These blocks are ranked according to their a.c. energy (a.c. energy quantifies the amount of information within a particular block). A bit assignment according to the average point variance within the block then takes place. It is here that the data compression takes place; more bits are assigned to visually "important" regions (i.e. regions of the image containing most information) than to those of lesser interest.

The Discrete Cosine Transform, F(k) of a discrete function f(j), j = 0,1, ..., N-1 where N is the set of data points is :

$$F(k) = \frac{2c(k)}{2} \sum_{j=0}^{N-1} f(j) \cos\left[\frac{(2j+1)k\pi}{2N}\right]$$

where k = 0, 1, ... , N-1 and $c(k) = \frac{1}{\sqrt{2}}$ for k = 0 and $c(k) = 1$ for k = 1, 2, ..., N-1

Previously, the DCT has been implemented using a double size Fast Fourier Transform (FFT) employing complex arithmetic and operating on 2N coefficients. The Fast Discrete Cosine Transform (FDCT) (Chen et al, 1977) alleviates the implementation problems associated with the DCT by using only real arithmetic and operating on N data points. This results in a factor of six reduction in the algorithm complexity.

The FDCT is most readily expressed in terms of an extensible flow graph. The 1-dimensional 8-point Fast Discrete Cosine Transform is shown in Figure 1.

4.3 Synthesis and Simulated Annealing

This section describes the behavioural synthesis procedure. The simulated annealing algorithm is introduced as a general purpose optimisation technique which has been applied most notably in VLSI floorplanning

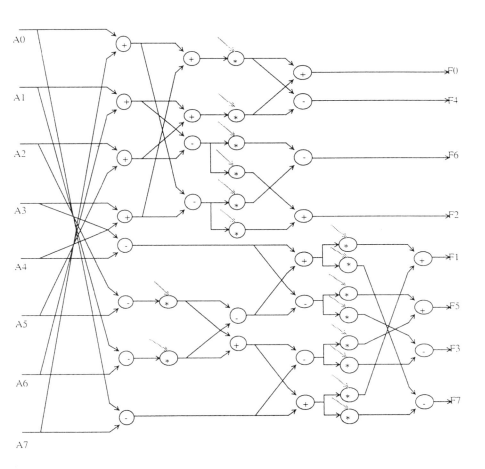

Figure 1 1-dimensional 8-point fast discrete cosine transform

problems. A formulation of the behavioural synthesis procedure is introduced which is amenable to a simulated annealing based implementation, and a relationship between the synthesis flow advocated by Denyer (Denyer, 1989) and the simulated annealing algorithm is developed.

4.3.1 The behavioural synthesis procedure

The behavioural synthesis task may be defined, at a high level, as the translation of a set of algorithmic descriptions of the required system behaviour into some suitable circuit formulation. This task may be subdivided as follows :

1) Compilation into a suitable intermediate data-structure. Current research concentrates on a relatively small core of data models, typically represented as either separated control and data flow graphs (SARI), combined control and data flow graphs (EASY (Stok and van de Born, 1988)) or tree structures (SILAGE (Hilfinger, 1984), Mimola (Marwedel, 1985)). Typical compiler optimization techniques can be applied at this stage.

2) Scheduling and Allocation. The scheduling subtask deals with the assignment of a suitable control step to individual data-flow graph operations, while the allocation subtask assigns particular data-flow graph operations to functional units. These subtasks are intimately related, for in order to determine an efficient schedule, some knowledge about the functional unit allocation is required, whilst allocation cannot take place without an indication of parallelism within the data flow graph, which, in turn, comes from the schedule.

3) Structural Synthesis. Within this step, the necessary memory and communications infrastructure required to complete the datapath, subject to the behavioural specification is generated.

4) Controller synthesis. This stage generates a suitable controller capable of sequencing data-flow operations on the specified datapath.

The version of SAVAGE reported here is a prototype system designed to investigate the scheduling and allocation stages of the behavioural synthesis procedure. Once a suitable schedule and allocation has been determined, then other software tools are invoked to complete the structural synthesis.

4.3.2 The simulated annealing algorithm

Simulated Annealing is a stochastic computational technique derived from statistical mechanics for finding near globally minimum cost solutions to large optimisation problems. Kirkpatrick, Gelatt and Vecchi (Kirkpatrick *et al*, 1983) were the first to propose and demonstrate the application of simulation techniques from statistical physics to problems of combinatorial optimisation, specifically to the problems of wire routing and component placement in VLSI design.

In general, finding the global minimum value of an objective function with many degrees of freedom subject to conflicting constraints is an NP-complete problem (Romeo and Sangiouanni-Vincentelli, 1985), since the objective function will tend to have many local minima. A procedure for solving hard optimisation problems should sample values of the objective function in such a way as to have a high probability of finding a near optimal solution and should also lend itself to efficient implementation. Recently, simulated annealing has emerged as a viable technique which meets these criteria. Rutenbar (Rutenbar, 1989) provides an elegant disposition on the subject.

The following pseudo-code function illustrates the structure of the subclass of probabilistic hill climbing (PHC) algorithms known as *simulated annealing*.

80 Simulated annealing based synthesis

```
function sim_anneal (initial_state, k0)

I       : STATE;
K       :CONTROL_PARAM;
COUNT   :INTEGER;

begin
  K = k0;
  I = initial_state;
  while (not stopping criterion) loop
    for count = 1 to #MOVES
      generate a new state;
      compute change in system energy, ΔE;
      if (ΔE < 0)
          /* LOWER COST - ACCEPT IT */
          accept this move; update I;
      else
          /* HIGHER COST - ACCEPT IT MAYBE */
          accept with probability P = e^−ΔE/K
          update I iff accepted;
    end for;
    update K;
  end while;
end sim_anneal;
```

where I is a state variable (in this case the datapath state), and K is the control parameter which models the temperature in the physical annealing system.

ΔE represents the change in energy between the current state and the state produced by the random perturbation of the data flow graph. The assessment of energy or cost, is discussed in Section 4.2.3. The stopping criterion is defined as $\Delta E = 0$ over 3 control parameter decrements. This ensures that the data flow graph has assumed a minimum energy configuration. The inner loop counter #MOVES determines the number of state generations per control parameter value.

In SAVAGE, the control parameter update function is defined :

$$K_{n+1} = K_n \, \alpha(K_n)$$

where $0 < \alpha K_n < 1$

4.3.2.1 Synthesis and simulated annealing

There exist a number of synthesis systems which use simulated annealing to produce data paths. Most notable are those developed by Devedas and Newton (Devedas and Newton, 1987 and Devedas and Newton, 1989), and Safir and Zavidovique (Safir and Zavidovique, 1990).

We can formulate the scheduling and allocation problem in terms of individual data flow graph node placement within a Resource-time (*Rt*) space. Rt space can be viewed, at the simplest level, as a bounded grid whose axes represent the various hardware units available to execute data flow operations and machine execution cycles, or *c-steps*, respectively. The scheduling and allocation operations may then be defined as a node displacement in *Rt* space, subject to individual data flow graph dependencies.

We can develop a simulated annealing based synthesis model through the integration of the linear design flow (described in Section 4.1) into the generate function of the simulated annealing algorithm. Selecting the finest computational "grain" (i.e. operating on single data-flow graph nodes), we can ensure that hill climbing moves can be attained at a minimal global cost. Every state generation cycle selects a data-flow graph node at random from the node set, assigns a c-step value to it, and binds it to a particular hardware resource, as shown in Figure 2.

4.3.2.2 Scheduling and allocation move set development

The scheduling component of the node translation was partitioned into 3 main stages. For the selected node, the *valid schedule range* is computed first. This operation is shown in Figure 3, and represents the computation of the upper and lower bound on the temporal displacement. A sequence of possible execution times is then randomly generated within the valid schedule range. The length of this sequence is proportional to the size of the valid schedule range. Finally, an execution time is selected at random from this sequence. This corresponds to the new execution time of the node.

By using this technique, a number of refinements were added to the basic scheduling operation. Selection of an execution time generated at random over the total valid schedule range ensured that genuine hill climbing moves were made available to the annealing procedure. The adaptive nature of the length of the execution time sequence increases the efficiency of the algorithm towards the end of the annealing run, where lower cost moves are generally achieved during the allocation phase, as most nodes have tended towards their optimum As-Soon-As-Possible (ASAP) schedule. Finally, to increase the performance of the scheduling algorithm during the early stages of the optimisation, where potential hill climbing moves have little effect on the overall quality of the final solution, the execution time sequence can be 'biased' to produce sequences of predominantly earlier execution times ($t_S' < t_S$) forcing a trend towards rapid ASAP type schedules.

The allocation move set developed was, by necessity, more deterministic in nature than the scheduling move set. In the most general view of synthesis, the module allocation procedure must ensure that a hardware component capable of executing the operation class is available at the scheduled time, t_S. A greedy heuristic allocation strategy will produce a module allocation equivalent to the maximum degree of parallelism of a particular operation class within a specific data flow graph. For practical purposes, this scheme represents a very inefficient use of available silicon area. Within SAVAGE, the allocation strategy is based on a 3 phase scheme. First, all hardware modules not supporting the operation class of the selected node are eliminated from the computation. From the remaining modules, a target hardware unit is selected based around a simple load balancing criteria. If the target module is free at t_S', then a simple binding between node and hardware module is established. If t_S' is unavailable on the target module, then it is eliminated from the set of candidate modules, and the allocation process reinvoked. Should the allocation process fail (i.e. all candidate modules are flagged as busy during t_S'),

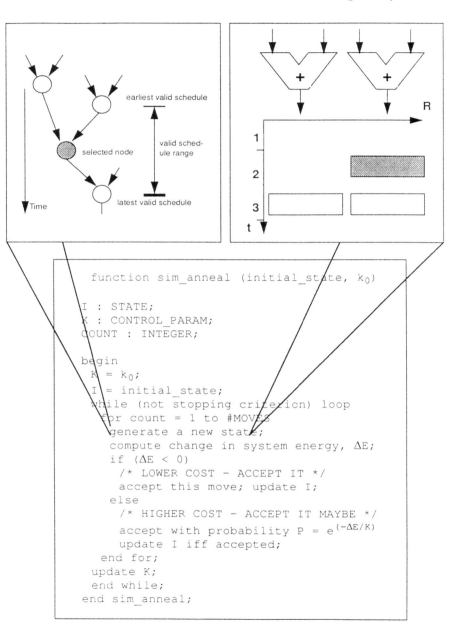

Figure 2 Integrating the scheduling and allocation into the annealing algorithm

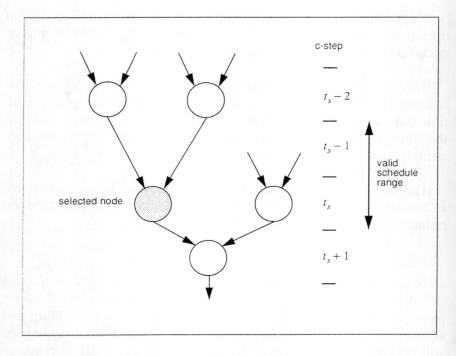

Figure 3 Computing the valid scheduling range

then in the earliest SAVAGE system, the user was prompted to either alter the valid schedule range of the node (i.e. manually alter the schedule), or allocate an extra hardware module of a corresponding class.

This manual intervention led to slow run times and a tendency towards greedy module allocation. Subsequently, the allocation strategy was revised to support operation deferment. The target module was still selected according to a load balancing criteria, but if the module was unavailable, then an extra c-step was inserted at the appropriate execution time, and the binding process invoked.

This allocation strategy allows a minimal hardware set to be used when operating under a time constraint, and ensures a global balancing of operation concurrency during the allocation phase.

SAVAGE supports a simple pipelining algorithm, similar to that described by Mallon (Mallon and Denyer, 1990). Here, the pipeline reuse time (initiation interval of successive pipeline tasks) may be specified as the timing constraint, and the pipeline latency (input to output latency of a single pipeline task) is optimised.

This pipelining operation can be viewed as a "folding" of the *Rt* space so that operations occurring after the computed pipeline reuse time are retimed to occur in free cycles in the next pipeline task.

4.3.2.3 Datapath costing

In developing a costing method for SAVAGE, a number of factors have to be considered. Firstly, as SAVAGE operates only on an incomplete part of the datapath solution space (namely the scheduling and allocation phases of the synthesis procedures), the costing functions used will not reflect the true cost of the datapath. The SAVAGE operational scenario has the user constraining one axis of the Resource-time space before the optimisation procedures are invoked. Correspondingly, the primary element of the costing function has to assess whether the datapath generated lies outwith the axis boundary specified by the user. Designs violating these boundaries are penalised heavily.

Part of the design specification for the SAVAGE software was to achieve a high utilisation of the functional units used within the solution datapath. The costing functions reflect this by penalising those designs which have functional units operating below a specific utilisation threshold (also specified by the user).

Further, datapaths generated as a result of more global perturbations to the solution - for example, where an extra control step is inserted into the schedule, and subsequent operations are retimed - are also penalised.

Thus the costing mechanism may viewed as a hierarchical structure, where gross system objectives are assigned a high importance while strategic goals, such as the attainment of a minimum functional unit utilisation occur at a lower level in the cost assessment hierarchy. Finally, library specific costing functions occur at the lowest level.

In this way, the annealing procedure is guided towards a solution which satisfies the gross system requirements quickly.

4.3.2.4 Costing mechanics

In common with Devedas and Newton, we formulate the datapath cost as a weighted sum of all hardware components within the datapath, combined with a weighted cost accounting for the total number of c-steps needed. (In the prototype SAVAGE system, structural synthesis takes place after the scheduling and allocation operations had been completed, and so the costings associated with these components were unavailable to the simulated annealing procedure.)

We extend the Devedas and Newton costing in keeping with the hierarchical costing model described above. Thus :

$$
\begin{aligned}
\text{COST}_{\text{DATAPATH}} = \;& W_1.\text{VIOLATIONS}_{\text{BOUNDARY}} \\
+\;& W_2.\text{VIOLATIONS}_{\text{FU\_UTILISATION}} \\
+\;& W_3.\text{VIOLATIONS}_{\text{RETIMING}} \\
+\;& W_4.\text{\#FUNCTIONAL\_UNITS} \\
+\;& W_5.\text{\#C-STEPS}
\end{aligned}
$$

The weightings can be varied to produce datapaths of varying architectural styles. For example, where the designer does nor explicitly wish to constrain the hardware resources available, but would prefer a solution with only a single multiplier, then the multiplier weight can be set proportionally higher, so that single multiplier solutions will have a lower global cost.

4.4 *Test results*

SAVAGE operates in a batch mode with the designer constraining either the hardware set available or the overall execution time desired.

(As SAVAGE can support simple pipelining, then the pipeline reuse time can be specified as a timing constraint)

The 1-dimensional 8-point FDCT was coded in SLANG (the SARI input LANGuage) as shown in Figure 4. The resulting data flow graph corresponds to Figure 1. The results shown in Table 1 were produced by specifying a hardware set for SAVAGE apart from those indicated as a pipelined solution, where a specific pipeline reuse time was specified.

Cycles	+	×	-	Av FU Util	Pipeline Reuse	Prop. Delay
20	1	1	1	70%	20	20
11	2	2	2	63%	11	11
8^1	2	2	2	87%	8	15
8	3	3	3	58%	8	8
6^1	3	3	3	77%	6	9
1. Pipelined execution plan						

Table 1 SAVAGE test results

A metric commonly used when assessing digital systems is the utilisation of each functional unit within the system. Here, it can be seen that the pipelined solution with a reuse time of 8 cycles offers the best time/hardware trade-off, and so this partially completed datapath was selected as the target for the remaining structural synthesis.

4.4.1 The structural synthesis tools

The prototype SAVAGE system used a simple left-edge algorithm to produce the memories required for the computed results and intermediate signals within the data flow graph. This algorithm produces the optimal memory *allocation*, but does not produce the optimal signal *groupings*.

88 Simulated annealing based synthesis

```
procedure FDCT_1D        ( A0,A1,A2,A3,A4,A4,A6,A7 : in FLOAT;
                           F0,F1,F2,F3,F4,F5,F6,F7 : out FLOAT) is

      COS_PI_4   : constant    :=  0.707_106_78;      -- cos(PI/4)
      COS_PI_8   : constant    :=  0.923_879_53;      -- cos(PI/8)
      SIN_PI_8   : constant    :=  0.382_683_43;      -- sin(PI/8)

      COS_3_PI_16 :constant    :=  0.831_469_61;      -- cos(3*PI/16)
      SIN_3_PI_16 :constant    :=  0.555_570_23;      -- sin(3*PI/16)

      COS_5_PI_16 :constant    :=  SIN_3_PI_16;       -- cos(5*PI/16)
      SIN_5_PI_16 :constant    :=  COS_3_PI_16;       -- sin(5*PI/16)

      COS_7_PI_16 :constant    :=  SIN_PI_16;         -- cos(7*PI/16)
      SIN_7_PI_16 :constant    :=  COS_PI_16;         -- sin(7*PI/16)

      B0,B1,B2,B3,B4,B5,B6,B7 : FLOAT;
      C0,C1,C2,C3,C4,C5,C6,C7 : FLOAT;
      D0,D1,D2,D3,D4,D5,D6,D7 : FLOAT;

      COS_PI_4_TIMES_B5 : FLOAT;
      COS_PI_4_TIMES_B6 : FLOAT;

      COS_PI_4_TIMES_D0 : FLOAT;
      COS_PI_4_TIMES_D1 :.FLOAT;

begin

      B0 := A7 + A0; B1 := A6 + A1;                   -- first pass
      B2 := A5 + A2; B3 := A4 + A3;
      B4 := A3 - A4; B5 := A2 - A5;
      B6 := A1 - A6; B7 := A0 - A7;

      -- Put the expressions COS_PI_4*B5 and COS_PI_4*B6 into intermediate
      -- variables so as to avoid evaluating them twice

      COS_PI_4_TIMES_B5 := COS_PI_4*B5;               -- second pass
      COS_PI_4_TIMES_B6 := COS_PI_4*B6;

      C0 := B3 + B0; C1 := B2 + B1;
      C2 := B1 - B2; C3 := B0 - B3;
      C4 := B4;
      C5 := COS_PI_4_TIMES_B6 - COS_PI_4_TIMES_B5;
      C6 := COS_PI_4_TIMES_B6 + COS_PI_4_TIMES_B5;
      C7 := B7;

      D0 := C0; D1 := C1;                             -- third pass
      D2 := C2; D3 := C3;
      D4 := C4 + C5; D5 := C4 - C5;
      D6 := C7 - C6; D7 := C7 + C6;

      -- Put the expressions COS_PI_4*D0 and COS_PI_4*D1 into intermediate
      -- variables so as to avoid evaluating them twice

      COS_PI_4_TIMES_D0 := COS_PI_4*D0;
      COS_PI_4_TIMES_D1 := COS_PI_4*D1;

      F0 := COS_PI_4_TIMES_D0 + COS_PI_4_TIMES_D1;    -- fourth pass
      F4 := COS_PI_4_TIMES_D0 - COS_PI_4_TIMES_D1;
      F2 := SIN_PI_8*D2 + COS_PI_8*D3;
      F6 := COS_3_PI_16*D3 - SIN_PI_8*D2;
      F1 := SIN_PI_16*D4 + COS_PI_16*D7;
      F5 := SIN_5_PI_16*D5 + COS_5_PI_16*D6;
      F3 := COS_3_PI_16*D6 - SIN_3_PI_16*D5;
      F7 := COS_7_PI_16*D7 - SIN_7_PI_16*D4;

end FDCT_1D;
```

Figure 4   SLANG description of FDCT

A greedy bus merger algorithm was used to synthesise the communications infrastructure required to complete the datapath. Here, replicated links between functional units (including the newly synthesised memories) are removed.

SAVAGE has been coupled with other datapath synthesis tools (Neil and Denyer, 1990) to synthesise a 5th Order Wave Digital Filter. Later iterations of the SAVAGE software include a complete route to datapath synthesis where structural synthesis is examined more fully.

4.5 Conclusions

This Chapter has described SAVAGE, a software tool capable of synthesising datapaths from behavioural descriptions. SAVAGE has been used in the development of datapaths for the 1-dimensional 8-point Fast Discrete Cosine Transform.

We have shown that given either a speed or hardware bound, SAVAGE can produce optimised solutions for both pipelined and non-pipelined designs which are comparable with those in published literature(Mallon and Denyer, 1989). SAVAGE allows the designer to rapidly explore the solution space for a given problem, and by varying the optimisation criteria produce a number of comparable datapaths. The designers own expertise is then used to select the most appropriate datapath solution.

SAVAGE has also been used to develop solutions to other synthesis benchmarks, notably the 5th Order Elliptic Wave Digital Filter, popularised by Paulin (Paulin and Knight, 1989 and Neil and Denyer, 1990).

4.5.1 Current developments

The software architecture for the SAVAGE toolset has been shown to be robust and flexible during the design cycle. Further structural synthesis move sets have been added to complete the SAVAGE route to datapath generation. This has been complemented by a corresponding increase in complexity of the datapath costing function. Indeed, move sets have been added which support a number of different architectural styles. This expansion has led to the development of eXtended SAVAGE (XSAVAGE); this CAD tool supports the "Architectural Script" based synthesis paradigm, first introduced by De Man (DeMan, 1990 and DeMan October, 1990). XSAVAGE is characterised by a 4 level hierarchy of user interaction, namely :

1) *System Level Interaction.* This level of interaction enables us to convey system level information, such as total chip area, maximum acceptable power consumption and timing specifications to the optimisation system.

2) *Strategic Interaction.* At this level in the hierarchy, we can specify the optimisation techniques that will form the generate function within the simulated annealing core. These comprise scheduling, allocation and memory and communication synthesis strategies. Also included here is the costing information.

3) *Pragmatic Interaction.* In many synthesis systems, application specific designer knowledge cannot be included in the specification. We provide a mechanism by which designers can affect the synthesis procedures directly via architectural pragmas.

4) *Structural Interaction.* At the lowest level in the script hierarchy, interaction takes place at the component level. Partial and complete architectures can be specified through the SAVAGE Structural Description Language.

4.5.2 Concluding remarks

The previous section describes XSAVAGE, a software tool which has evolved from the SAVAGE software which was initially intended to synthesise Fast Discrete Cosine Transform blocks. XSAVAGE is a much more powerful system, capable of producing optimised solutions of widely differing architectural styles for a given problem domain. XSAVAGE may be classified not as a *problem-specific* synthesis system, but rather as a general synthesis *framework* capable of supporting application specific architectures.

4.6 Acknowledgements

This work was carried out as part of the Silicon Architectures Research Initiative. The use of facilities and resources is gratefully acknowledged. The work reported here is supported by the Science and Engineering Research Council and the University Of Edinburgh.

4.7 References

CHEN W., SMITH C.H. AND FRALICK S.C., "A Fast Computational Algorithm for the Discrete Cosine Transform", IEEE Trans. Commun., 1977, com25, (9), pp. 1004-1009.

CHEN W. AND SMITH C.H., "Adaptive Coding of Monochromatic and Colour Images ", IEEE Trans. Commum., 1977, com25, (11), pp. 1285-1292.

DE MAN H., "Tutorial On High-Level Synthesis" EDAC '90, March 1990.

DE MAN H., "CAD For Real Time Information Processing Systems : Challenges and Opportunities", in Proc. SASIMI '90, October 1990, Kyoto, Japan, pp. 65-72

DENYER P.B., "SAGE Design Methodology", SARI Internal Technical Report SARI-035-D, March 1989.

DEVEDAS S. AND NEWTON A.R., "Algorithms for Hardware Allocation in Data Path Synthesis," in Proc. ICCAD '87, 1987, pp. 526-531.

DEVEDAS S. AND NEWTON A.R., "Algorithms for Hardware Allocation in Data Path Synthesis," IEEE Trans. Computer-Aided Design, Vol. CAD-8, No. 7, July1989, pp. 768-781.

GRANT P.M., "The DTI-Industry Sponsored Silicon Architectures Research Initiative", IEE Electronics & Communications Engineering Journal, Vol. 2 No. 3, June 1990.

HILFINGER P.N., "SILAGE: A Language for Signal Processing", University of California, Berkley, 1984.

KIRKPATRICK S., GELATT C. AND VECCHI M., "Optimisation by Simulated Annealing," Science: Vol. 220, No. 4598> May, 1983, pp671-680.

MALLON D. AND DENYER P.B., "Behavioural Synthesis : An Interactive Approach," IEE Colloquium Digest 1989/85, May 1989, pp.2/1-2/8

MALLON D., AND DENYER P.B., "A new Approach to Pipelining Optimisation", in Proc. European Design Automat. Conf. , March 1990.

MARWEDEL P., "The MIMOLA Design System: A Design System which spans several levels", in Methodologies of Computer System Design, ed. Shriver, North Holland, 1985, pp. 223-237.

NEIL J.P. AND DENYER P.B., "Exploring Design Space using SAVAGE: A Simulated Annealing based VLSI Architecture GEnerator", in Proc. 33rd Midwest Symposium on Circuits and Systems, Calgary, August 1990.

PAULIN P.G. AND KNIGHT J.P., "Scheduling and Binding Algorithms for High-Level Synthesis," in Proc. 26th Design Automat. Conf., June 1989.

ROMEO F. AND SANGIOVANNI-VINCENTELLI A., "Probabilistic Hill Climbing Algorithms : Properties and Applications," in Proc. Chapel-Hill Conf. on VLSI, 1985.

RUTENBAR R.A., "Simulated Annealing Algorithms : An Overview," IEEE Circuits and Devices Magazine, January 1989, pp. 19-26.

SAFIR A. AND ZAVIDOVIQUE B.,"Towards a Global Solution to High Level Synthesis Problems", in Proc. European Design Automat. Conf., March 1990, pp. 283-288.

SOAME T.A., "Bandwidth Compression of Images Using Transform Techniques", GEC J. Sci. Tech., 1982, 48, (1), pp. 17-23.

STOK L. AND VAN DER BORN R., "EASY: Multiprocessor Architecture Optimisation", in Proc. Int. Workshop on Logic and Architecture Synthesis for Silicon Compilers, Grenoble, May 1988, pp. 313-328.

WINTZ P.A., "Transform Picture Coding", Proc. IEEE, 1972, 60, (7), pp. 809-819.

Chapter 5
Knowledge Based Expert Systems in Testing and Design for Testability - An Overview.

G. Russell

5.1 Introduction

Over the past decade CAD tools to assist in the conceptual design and layout phases of a VLSI circuit have reached a high degree of sophistication, enabling engineers to be more creative in their designs by removing the more tedious aspects of the design process. The result of this design methodology has, unfortunately, been the rapid production of complex circuits which cannot be tested economically. In the past, the testing phase of a design has been considered as the pariah of the design process and thus not worthy of consideration by the designers. However, the problem of testing VLSI circuits is a major obstacle to the pervasive use of this technology. It must be accepted that solving the problem of testing a circuit is as much part of the design process as designing the layout of the circuit or verifying the functional correctness of the circuit and must be considered early in the design cycle and not left as an after thought when the design is completed.

The symptoms, in general, of the problems of testing VLSI circuits (Eichelberger and Lindbloom, 1983 and Wilson and Hnatek, 1984) are:-

1) Increase in testing costs, which are proportional to testing time and increase with circuit complexity.

2) Increase in test generation and evaluation time.

3) Increase in the volume of test data.

These problems of increased testing costs, overall, are particularly pertinent to the ASIC market since,

1) ASICs require fast design times if the products in which they are to be used are to retain their competitive edge. Subsequently they have a short lifetime.

2) Due to the unique functions performed by ASICs, production volumes are low and test costs cannot be amortised over a large number of devices.

3) The uniqueness of ASIC functions requires new test programs to be developed for each design.

In an attempt to reduce the cost of ASICs users have been designing the circuits themselves; having no appreciation of the problems of testing complex circuits, they tend to aggravate test problem by generating inefficient test sets and setting unattainable test objectives.

Attempts to reduce testing costs, in general, have been made by:-

1) Developing more sophisticated test generation algorithms.

2) Incorporating Design for Testability techniques on the chip.

Although new test generation algorithms can produce sets of test vectors efficiently, the length of the test sequences generated are greater than those produced by test engineers.The fundamental reason for the difference is that the test generation algorithms operate at too low level of abstraction without regard to the semantics of the functions in the circuit. Furthermore, each circuit is considered as a unique entity. The test engineer, however, not only views the circuit at a higher level of abstraction, but also uses 'knowledge' accumulated from past experiences when developing a test set for a given circuit.

Design for Testability techniques have been shown to reduce testing costs, but unfortunately the expertise required to design testable circuits is not widely available. Consequently, when a designer is faced with the problems of designing a testable circuit, he is not aware of all the possible techniques or the ramifications of a particular method with respect to his design. Hence a design is produced which has improved testability, but incurring high overheads.

The necessity to use 'knowledge' or 'expertise' gained from past experiences to produce either an efficient test program or a testable design, has motivated the development of a new generation of Computer Aided Test (CAT) tools. These new CAT tools incorporate AI techniques which permit 'expert' knowledge to be embodied within a tool so that it can not only generate 'expert' solutions to given problems but also 'explain' and 'justify' why a given solution has been proposed.

However, the inference that certain CAD programs are 'intelligent', 'expert' or 'can think' generates an unwarranted mystique about them, and many definitions of expert systems, unfortunately, tend to engender and support this aura of mystery. Part of the mystique about expert systems has arisen from the type of problems that these programs have been used to solve, typically they are problems which cannot be solved algorithmically and have only been solved in the past through the use of human expertise.

To highlight the developments of expert systems in test generation and design for testability, several examples will be described. However, before describing the applications a brief description of the components of an expert system will be given.

5.2 Components of a Knowledge Based Expert System

(Hayes-Roth, 1984 and Waterman, 1986)

Briefly, an expert system is a knowledge intensive program capable of producing solutions to a range of non-numerical problems, which in the past required the knowledge of a human expert to provide the solution. An expert system differs from conventional programs in that it operates on knowledge in the form of 'rules and facts' rather than data which is passive; the procedures invoked tend to be heuristic, although not necessarily so, rather than algorithmic, and solutions are obtained by an inferential process rather than some repetitive numerical process. In general, a conventional program does not contain explicit knowledge of the problem to be solved, it simply contains a sequence of instructions, defined by the programmer, to solve a specific problem. Although there is much debate whether certain programs can be considered to be expert systems or algorithms, a broadly accepted characteristic of an expert system is that the control or procedural reasoning mechanism is separate from the declared knowledge. The programming environment provides the 'control' by means of a general purpose 'inference engine' which can reason about facts regardless of the domain of application. The body of facts or operative knowledge, ie the 'expertise' about a given problem domain is contained,usually, in the knowledge base. The accumulation and representation of the body of expertise is one of the most important aspects of an expert system.

An expert system comprises, essentially , four modules, namely a Knowledge Base, an Inference Engine, a Knowledge Acquisition Module and an Explanatory/Interface Module.

The Knowledge Base contains all the information required by an expert system to make it act 'intelligently'. The knowledge in the system comprises facts and rules, which use the facts to make decisions; 'certainty' factors may also be associated with the facts and rules to indicate the degree of confidence that a given fact or rule is true or valid. Some of the rules in the knowledge base may also be heuristic, that is, 'rules of thumb' which ,for example, in the interest of efficiency will limit the area of search for a possible solution and hence will not guarantee to find the solution to a given problem, although in most cases a satisfactory solution will be found. Within an expert system the knowledge can be represented and organised in several ways. The majority of systems tend to be knowledge based, that is, the knowledge about the problem to which the system is to be applied is separate from the other modules in the system, this has the advantage of permitting fast prototyping of systems together with the ease of adding new facts or rules to the knowledge base. Within the knowledge base the rules and facts can be structured in a number of ways. Several of the more common methods(Waterman, 1986) used are Ruled-Based, Logic-Based and Frame-Based. In the ruled-based method, knowledge is represented by antecedent-consequent rules of the form 'IF < Condition>THEN < Action >'; the antecedents or conditions may be joined by logical connectives, the consequent or action in a rule is only performed if the conditions in the problem being solved match the antecedents in the rule. The logic-based technique for knowledge representation uses a logic programming language to write declarative clauses to describe facts, rules and goals. The general format of a clause is 'consequent:- antecedent 1,antecedent 2...............etc.'. A clause without an antecedent is a fact, a clause without a consequent is a goal or objective to be validated and a clause containing both consequent and antecedents is a rule. The most commonly used language for logic-based systems is Prolog(Clocksin and Mellish, 1981). The frame-based system comprises a data structure with a hierarchical network of nodes (frames). The topmost nodes representing general concepts or objects, the frames further down the hierarchy represent more specific instances of the concepts or objects. Within a frame declarative or procedural knowledge is stored in 'slots', each slot being assigned to a particular attribute of an object. For example, the frame for a flip-flop may contain slots for the 'type', 'technology','fanout', 'speed' etc., for a specific instance of a flip-flop these may contain the values 'J-K', 'CMOS', '8', '10ns'.The procedural information contained in a slot may be used activate some standard operations. The interconnections between the frames are used to express relationships between them.

Within the expert system the ability to 'reason' about facts in the knowledge base is achieved through a program called an Inference Engine. The most common 'reasoning' strategies used in an Inference Engine are the 'forward' and 'backward' chaining techniques. The forward chaining technique, sometimes called the data-driven technique, starts with the given facts and attempts to find all the conclusions which can be derived from the facts.The backward chaining technique essentially hypothesises about a given consequence and attempts to find the facts to substantiate it. The technique used depends very much upon the problem to be solved. The inference engine may be either explicit or implicit to the expert system, trading off the flexibility and subsequent improvement in efficiency in choosing the most appropriate 'reasoning' technique to solve a problem against minimising the amount of work required to produce an expert system. One of the disadvantages of using expert system shells, that is, an expert system without the problem domain knowledge, is that the inference mechanism may not the most appropriate 'reasoning' strategy for the particular problem domain.

The knowledge acquisition module is part of the knowledge-based system building aids that help to acquire and represent knowledge from a human expert in a given problem domain. The task of acquiring the knowledge from the human expert and choosing the most appropriate representation for it is the the function of the knowledge engineer. Several methods are available for extracting knowledge from the human(domain) expert which is one of the major bottle-necks in developing an expert system; some of the techniques used are on-site observation, problem discussion and problem analysis.The difficulties in generating a knowledge base are further compounded if more than one domain expert is consulted. Several tools have been developed to assist in generating a knowledge base, for example, in addition to the basic editing facilities there are syntax checkers to verify the grammar used in the rules, consistency checkers to ensure that the semantics of the rules or data are not inconsistent with knowledge already present in the system. There may also be a 'knowledge extractor' which assists the end-user of the system to add and modify rules, this is essentially a combined syntax and consistency checker together with prompting and explanation facilities should the end-user get into difficulties.

The Explanatory/Interface module is the means by which the end-user interacts with the system, although to date full natural language processing, excepted in some strict technical contexts, is unattainable. An important characteristic of an expert system is its ability to 'explain' their 'reasons' for coming to a given conclusion, this is not only useful in debugging prototype systems but also gives the end-user a degree of confidence in the system. Most expert systems have a 'predictive' modelling capability, that is, the user can interrogate the system to find out the effect of certain changes in a given situation or if the situation changed what caused the change.

In comparison to the human expert, the expert system offers the following advantages(Waterman, 1986),

1) In an expert system, the knowledge is permanent, that is facts are not forgotten through lack of use. The human expert will forget facts if they are not used frequently.

2) The 'reasoning' power of the expert is not subject to emotional factors, for example, stress which can make the human expert unpredictable.

3) The knowledge in the expert is easily transferred or reproduced, this is not the case with the human expert.

4) An expert system is affordable, a human expert can be extremely expensive.

A limiting factor regarding the efficiency of present day expert systems is that they are run on computer architectures which are entirely unsuitable. Expert systems and in general AI systems are concerned with acquisition, representation and the intelligent use of knowledge and employ symbolic rather than numerical processes in these functions. For example, in the acquisition of knowledge, which may be incomplete, imprecise, contradictory and come several sources, symbolic processing is required for the correct recognition and understanding of the input data, this involves the basic processes of comparison, selection, sorting, pattern matching, and logical set operations. Knowledge representation involves encoding information about objects, relations, goals, actions and processing these into appropriate data structures. The actual processing of knowledge for problem solving and making logical deductions requires efficient 'search' procedures. The efficiency of the search procedures reflects overall on the ability of the systems to produce 'expert' answers to given problems since it can traverse the solution space more effectively.

A comparison(Ramamoorthy *et al*, 1987) between conventional and AI type programs is shown below :-

FEATURE	CONVENTIONAL	AI
Type of Processing	Numeric	Symbolic
Techniques	Algorithmic	Heuristic Search
Definition of Solution Steps	Precise	Not explicit
Answer Sought	Optimal	Satisfactory
Control/Data Separation	Intermingled	Separate
Knowledge	Precise	Imprecise
Modification	Rare	Frequent

It has been suggested that 'Data-Flow' machines are better suited to the processing requirements of expert system programs, particularly 'rule-based' schemes, than conventional machines, since the rules can be represented as a flowgraph which is the natural programming structure for data-flow computers. Furthermore, data-flow machines handle parallelism automatically, permitting more efficient search procedures to be implemented.

5.3 Test Generation (Robinson, 1984 and Angwin et al, 1984)

Over the past decade improvements in the fabrication technology have permitted more complex systems to be integrated onto a single silicon chip . Although this accrued many advantages, it has created a problem which has become a major design issue, namely testing the circuits immediately after fabrication and also maintaining systems in the field. Many developments have been made to improve the efficiency of test generation algorithms, which has also been enhanced by the introduction of design for testability techniques; although these developments are reducing testing costs, the techniques have many limitations.The fundamental problem with the current approaches to the test generation problem is that tests are generated at too low a level of abstraction, namely gate level, where the semantics of the circuit are lost in the complexity ; at this level test generation is a very CPU intensive process. In comparison, the test engineer can develop effective tests for complex systems in a relatively short time. The reason for his success is that he views the circuit at a higher level of abstraction, namely functional level. Furthermore in developing tests he uses a wealth of knowledge obtained from the circuit under test and also from past experience. For example, he 'knows' what tests are effective in testing particular functions, or how to get the system into a given state, or how to control/observe the main interfaces between the high level functional blocks. In addition, he attempts to generate tests using as many of the normal functions of the circuit as possible. Consequently, it has been recognised that the generation of tests for VLSI circuits or systems is a highly knowledge intensive task, full of heuristics and not simply an algorithmic exercise carried on a gate level description of the circuit. Consequently this is an area where expert systems can be used to good advantage by incorporating some of the 'knowledge', used by test engineers, into test generation programs. Several expert systems have been developed for generating tests for VLSI circuits and diagnosing faults in digital systems, some of these will be described, briefly, below.

SUPERCAT(Bellon *et al*, 1983): This expert system was developed to assist designers to generate global test patterns for complex VLSI circuits. Automatic test pattern generator programs work efficiently on small circuits or cells, but the basic problem is generating the global test program for the complex circuits in which the basic cells are embedded, that is, deriving the primary input signals to activate the fault and propagate the fault information from the cell to an observable output.

The knowledge base in SUPERCAT contains information on,

1) A range of ATPG programs, for example, the types of circuits to which they can be applied, the types of faults considered, the format of the input description of the circuit used by the test generation program.

2) A library of commonly used test patterns which have been shown to be effective in the past (equivalent to test engineers experience).

3) Design for testability techniques which facilitate the generation of tests.

SUPERCAT was developed to operate on data driven and control driven circuits, each type of circuit is represented by a dataflow and control flow model respectively. When processing data driven circuits, the smallest group of functions which can be activated independently of the remainder of the circuit are identified, in this way signal paths required to control/observe particular functions can be determined. Thereafter the order in which the functions are to be tested is determined. A testability evaluation of the circuit is also performed to identify areas of the circuit where testability hardware should be located if more precise information is required on the location of faults. When performing the above functions SUPERCAT invokes the CATA(Robach *et al*, 1984) system.

In generating tests for control circuits, the circuit is first divided into two parts, namely, the controller and controlled resource and a far as possible normal circuit operations are used as the basis for the test patterns.

The essence of SUPERCAT is to suggest the most appropriate techniques to test a given function block from the data in its knowledge base and the information derived from the circuit and subsequently, within the restrictions imposed by the operation of the circuit, attempt to generate the necessary primary input signals to activate the test at the inputs of the function under test and propagate the output response of the function to a primary output.

HITEST(Parry *et al*, 1986, Robinson, 1983 and Bending, 1984):This is a test generation system which utilises a combination of algorithmic and knowledge based techniques to derive effective tests vectors for VLSI circuits. It has been recognised that a major limitation of current ATPG methods is that they view the circuit on a microscopic level, in terms of gates and flip-flops. Consequently as circuit complexity increase, test generation times increase, the volume of test data increases and test application times increase. Conversely test engineers can develop efficient test programs for complex circuits in a relatively short time. The reason for this is that the test engineer and the ATPG program are being asked to solve different problems. The test engineer has a more macroscopic view of the circuit and hence the test generation problem is more tractable; if the test engineer is only given a gate level description, test generation times would greatly exceed those of the ATPG program. Although several high level test generation programs have been developed to improve the efficiency of the test generation process these are still not as effective as the test engineer, since in addition to having a macroscopic view of the circuit the test engineer has a wealth of knowledge of past experiences, that he can use in testing a circuit, which does not exist in the test generation program. It is the objective of HITEST to integrate this body of test knowledge with algorithmic test generation methods to improve the efficiency of the automatic test generation process.The knowledge base used in HITEST is a frame-based system in which all the data is stored in slots contained within a frame. The knowledge which is stored comprises, for example, data about,

1) Parts of the circuit which are sequential so that the backtracker in the test generator can avoid them.

2) Techniques to control/observe well known types of signals.

3) Overall test strategies, that is whether to apply functional or algorithmically generated tests.

4) For a given circuit, the actual functions used from a multi-function block extracted from a cell library, since it is only necessary to test the actual functions of the block used by the circuit.

The knowledge stored in the system is directed at constraining the solutions proposed by the system to test sequences which,

1) Would be considered by a test engineer to be effective, if he had been asked to perform the task.

2) Could be translated, readily, into appropriate formats to drive a range of ATE equipment and make full use of the pattern generator facilities offered by the equipment.

HITEST has been used effectively on circuits comprising 3500 gates using 30 minutes of CPU time on a VAX11/750 giving 98% fault coverage.

SCIRTSS(Hill and Huey, 1977):There has been much debate about whether or not programs which incorporate heuristics to limit the scope of search procedures are expert systems. SCIRTSS (Sequential Circuit Test Search System) is a program which falls into this category. SCIRTSS is used to generate tests for faults in sequential circuits and incorporates several heuristics which limits its search space for a solution and also improve the probability of following the correct search path.

SCIRTSS uses two representation of the circuit during test generation, first, a high level description in AHPL which facilitates the search for the appropriate transition sequences during test generation and which overcomes the basic problem with gate level ATPG programs of losing the global view of the circuit behaviour. Second, a gate level description which is used to verify that the test sequence generated will detect the fault.

In generating a test it is assumed initially that the circuit is in a known state and SCIRTSS invokes the D-Algorithm to find a partial test for the given fault. In generating a partial test the D-Algorithm views the circuit as combinational by limiting its consideration of the circuit behaviour to one clock period. The test for the fault is propagated and justified to the circuit boundaries which are considered to be either primary inputs/outputs or memory elements. If the D-Algorithm procedures terminate on memory elements, two search procedures are subsequently invoked to complete the test sequence generation.

If it was the justification procedure which terminated on a set of memory elements, then a 'sensitisation' search is invoked to determine the sequence of inputs necessary to switch the circuit from its current state to the objective state defined by the justification process which will propagate the fault to a primary output or another set of memory elements.The AHPL description of the circuit implicitly specifies the state transitions and the conditions necessary for these to occur in the circuit.

The search for the necessary sequences is essentially a 'tree' search. If the fault propagation phase of the D-Algorithm terminated on a set of memory elements, then the 'propagation' search procedure is invoked to determine the necessary input sequences to make the fault visible.

During the search procedures inputs are selected heuristically and simulation at RTL level is used in searching the control state graph for the required sequences. The search procedures are continued until the required sequences are found or the number attempts made exceeds that specified by the user. During the search procedures, the process is controlled by heuristic values assigned to nodes in the circuit, calculated from user defined parameters. The heuristics, for example, limit the depth and width of the search, bias the search to control nodes which have not been used so far in the process.

When a test sequence has been generated a gate level simulation of the circuit is performed to determine if the test sequence will detect the fault and also to determine what other faults are detected by the test.

A number of knowledge based systems have also been developed for diagnosing faults in electronic equipment, for example,

1) MIND(Wilkinson, 1985), which is used to diagnose faults in VLSI testers. The system uses a hierarchical structural description of the tester together with 'expert' knowledge to reduce the down time of testers, which incur a loss of approximately $10,000 for each hour that they are out of service in a production environment.

2) CRIB(Hartley, 1986), which is used to assist engineers in the field to diagnose hardware/software faults in computer systems. The input to the expert system is an almost English language description of the symptoms of the fault, these symptoms are subsequently matched with known fault conditions in a database. The diagnostic expertise is represented as 'symptom patterns' which suggest actions to be performed to obtain further information on the symptoms of the fault.

3) FIS(Pipitone, 1987), which is used to assist technicians to diagnose faults in analogue electronic equipment; faults can be isolated down to the level of power supplies, amplifiers etc. The system has a novel feature of being able to reason qualitatively from a functional model of the system, without recourse to numerical simulation. The techniques employed in FIS can also used to diagnose faults in mechanical, hydraulic and optical systems.

Attempts are also being made to develop expert systems to assist in the diagnose of faults in analogue circuits, for example, the DEDALE (Deves et al, 1987) system and the work carried out by TERADYNE (Apfelbaum, 1986), the test equipment manufacturers.

5.4 Design for Testability

It has been established that the high costs of testing can be reduced by the use of Design for Testability (DFT) techniques. However the proliferation of DFT methods available, each with its own advantages/disadvantages has introduced an unfortunate side effect, namely the problem of choosing the best method for a given circuit. Again the characteristics of this problem are suited to solution by the use of an expert system techniques. A number of expert systems have been developed for the synthesis of DFT structures and comprise (Samad and Fortes, 1986), in general,

1) Options Generator, which proposes a number of solutions.

2) Cost Analyser, which compares the trade-offs of the different techniques proposed and may attempt to produce some qualitative figure of merit for each technique.

3) Synthesis system, which attempts to integrate the technique selected into the circuit, either using the hardware existing in the circuit or introducing additional function blocks into the design.

Some DFT systems also incorporate an 'Explanation' capability(Bounanno and Sciuto, 1988) in order to present the user with the reasons why a particular solution was proposed, since a number of possible solutions to a given problem may exist. Furthermore, during the development stages of the system it is important to know how the system achieved certain results or if there are any inadequacies in the Knowledge Base.

The type of knowledge stored in the Knowledge Base in a DFT system can be heuristic, procedural or structural.

1) Heuristic Knowledge - this is advice on what to do in given circumstances, this knowledge can be subdivided into :-

- Strategic knowledge, for example, knowledge concerned with choosing a DFT technique.

- Metalevel knowledge, which guides the problem solving process.

- Microlevel knowledge, which deals with particular steps in the problem solving process. For example, the order in which hardware modifications are made.

This knowledge is usually stored as 'rules'.

2) Procedural Knowledge - this is the sequence of actions necessary to achieve a particular objective. For example to implement a scan-path in a design it is necessary to replace all flip-flops with scannable latches, add some control logic etc.This knowledge is stored as a 'goal tree'.

3) Structural Knowledge - this includes the classification of circuit elements for, example all flip-flops are classified as memory elements, and causal models of behaviour which may be used to estimate area overheads, yield etc. This knowledge is stored as 'frames'.

As an example of expert systems developed to incorporate DFT structures into a circuit TDES(Abadir and Breuer, 1984), TIGER(Abadir, 1989), ADFT (Fung *et al*, 1985 and Fung and Hirschhorn, 1986), PLA-

ESS(Breuer and Zhu, 1985) and OPEC2(Geada and Russell, 1991) systems will be briefly described.

TDES(Testable Design Expert System)(Abadir and Breuer, 1984): The objective of TDES is to assist both in the implementation of a testable circuit and the production of a test plan for the circuit. This approach is referred to as a Testable Design Methodology(TDM). The input to TDES comprises a description of the circuit to be processed together with any constraints , for example, area overhead, pin out, ATE requirements etc. The knowledge base within TDES contains information on a range of DFT techniques, their attributes, rules concerning their implementation and a method of evaluating the trade-offs of the proposed solutions.Within TDES the circuit is represented by a hierarchical graph model, with the nodes representing function blocks and the arcs representing interconnections. Labels are assigned to each node which describe the attributes of the node, for example, number of inputs/outputs, number of product terms if it is a PLA etc. The label also specifies the function of the node, since TDES may use the structure represented by the node in the implementation of a DFT technique.

Since TDES also attempts to produce a test plan for the circuit , information, in the form of a template, is stored in the knowledge base on testing techniques, in terms of the functions to which the technique can be applied, the hardware necessary to generate the patterns, capture responses and evaluate the results, together with information on how to connect the hardware to the function (kernel) to be tested.The template also contains information on the trade-offs associated with a given method, so that the 'goodness' of an implementation can be evaluated, and the sequence of operations (test plan) necessary to test the kernel using a given DFT method.For the purposes of testing TDES divides the circuit into kernels, which may or may not be functional blocks, but have the salient feature that each kernel can be tested using a standard DFT technique. Each kernel is assigned a 'weight' by the system and the kernels with the highest weights are processed first, since these have been deemed the most difficult to test. The process of structuring a DFT technique around a kernel is called 'embedding' and comprises, essentially, matching the graph of the kernel to that of the TDM, if a direct match cannot be obtained additional hardware is added. If a kernel can be tested by several TDMs the designer is presented with the solutions and associated trade-offs and asked to select a solutions or reject all solutions.The final output of the system defines the techniques to be used on each kernel, the overheads incurred and the test-plan for the circuit giving a 'total' test solution.

The concepts developed in TDES have been refined and incorporated in the TIGER (Testability Insertion Guidance Expert) (Abadir, 1989) System. TIGER adopts a hierarchical testability synthesis approach which addresses the issues of DFT at block, module(RTL), and gate level.

TIGER accepts as input a hierarchical description of the circuit in either VHDL, EDIF or the internal TIGER language, together with the testability objectives and design constraints. The DFT knowledge in the system is contained in TDMs which store structural requirements, test-plans, cost functions etc. for the various DFT schemes. The first step in modifying a circuit to improve its testability in TIGER is to perform a high level testability analysis of the circuit to identify potential bottlenecks in the circuit and the location of probable test paths. This information, which would be difficult to obtain if only a gate level description of the circuit was available, is used by the partitioning and testability modules in TIGER. The next phase of the process is the partitioning of the circuit into primitive kernels, each of which has the property of being able to be tested by at least one of the DFT techniques stored in the knowledge base.However, to improve the overall solution regarding area overheads, attempts are made to 'cluster' the primitive kernels into larger kernels in order to obtain an optimised design. Once the clusters have been selected, the TDM matching and embedding process starts in which additional hardware is inserted into the circuit or existing structures modified so that a particular DFT method can be implemented for a given kernel.Thereafter the trade-off estimates are analysed and compared with the constraints. The constraints supplied by the user can be used to dynamically control a constraints-directed embedding process which dynamically adjusts its optimisation criteria in search for a feasible solution which satisfies the testability objectives without violating the constraints of the designer. After the embedding and evaluation processes are complete, the test-plan for the circuit is generated by TIGER which describes, symbolically, the sequences of actions necessary to test the various clusters/kernels in the circuit. The test-plan is also sent to the gate level test generator and fault grading modules in TIGER to generate the required test patterns and determine the fault coverage. These low level tools are only required to operate on individual clusters or kernels as directed by the test-plan and not the complete circuit, avoiding the complexity problems which normally render gate level tools inefficient. The final output from the TIGER system is a modified design file together with testability reports summarising the results.

ADFT(Automatic Design For Testability)System(Fung *et al*, 1985 and Fung and Hirschhorn, 1986):This systems differs from TDES in that it is integrated into a silicon compiler where the testability enhancements to be incorporated are synthesised with the functional hardware early in the design cycle.The objective of the system, overall, is not only to guarantee 'correctness by construction' but also 'testability by construction'.

The ADFT system comprises,

1) A set of rules to enforce design for testability together with a 'rules checker'.

2) A testability evaluator which identifies 'hard to test components'.

3) A library of testability structures.

4) TEXPERT, a testability expert system.

The testability evaluator determines the testability of a design based on cost criteria, design constraints etc. and subsequently passes information on testability bottlenecks to the testability expert system,TEXPERT, which suggests intelligent design modifications to alleviate the difficulty. The function of TEXPERT is to decide where and what testability enhancement structures have to be integrated into the circuit. Its decisions are based on information extracted from a number of sources, for example, the testability evaluator, the testability 'rules checker' test requirements/constraints imposed on the circuit by the designer etc. If a function block in the circuit is deemed 'hard to test' by the testability evaluator, an attempt is made to replace it with a more testable version from the system function library. All of the testability enhancement structures integrated into the circuit being designed are connected , for the purposes of testing, to a control bus system called OCTEbus(On Chip Test Enhancement bus) which guarantees over controllability/observability of all function blocks in the circuit regardless of hierarchy.

PLA-ESS(PLA-Expert System Synthesiser)(Breuer and Zhu, 1985): In general , traditional test methods are not suitable for PLAs due to their high fan-in/fanout characteristics. Consequently a number of DFT techniques have been designed to ease the problem of testing PLAs, each technique having its own special requirements and overheads. Thus the designer is faced with the problem of choosing a method from a range of possible solutions which is most suitable to his requirements; furthermore the designer may be unaware of certain solutions to a particular testing problem which are more suitable to his requirements. Consequently,in choosing the best method for a PLA, the designer must examine a multi-dimensional solution space in which global trade-offs have to be made. It is to assist in this task that PLA-ESS was developed.

The input to PLA-ESS consists of a description of the PLA together with a requirements vector which defines the restrictions to be imposed on the DFT technique with respect to area overhead, fault coverage, external test requirements etc. A weighting can be assigned to each attribute of the constraints vector to indicate its relative importance. After PLA-ESS has accepted the PLA description and requirements vector, an evaluation matrix is generated which defines the attributes of each DFT method in the knowledge base of PLA-ESS with respect to the PLA being processed.

A search of the evaluation matrix is then made to determine if any entry matches the requirements vector. If there is a unique solution it is identified. If several solutions exist, the best one is selected upon evaluation of a 'scoring' function. The 'scoring' function is a factor which reflects the trade-offs of each technique and its overall effect on the performance of the circuit, allowing the designer to make a quantitative judgement on the best technique. However, if a solution does not exist, the 'reasoner' is invoked to determine why a solution could not be found and how this situation may be resolved. This is usually achieved by modifying the constraints requirements so that they are less severe. The designer can seek advice from the system about which attribute of the requirements vector needs to be altered to obtain a solution. If several changes have been made, the designer can request a 'trace design history'. The designer is not obliged to accept any of the solutions proposed by the system, but can make his own decision and monitor its outcome. If a solution or suggestion is rejected the system will prompt for a reason and then invoke the 'reasoner' to determine if a more acceptable solution exists. Once a suitable technique has been found the PLA is modified so that the technique can be implemented and an appropriate test-set is generated.

OPEC2 (Optimal Placement of Error Checking Circuitry) (Geada and Russell, 1991): This is an automatic DFT synthesis system targeted on a self-testing method which uses coding (information redundancy) techniques for concurrent error detection. By being specifically targeted at one method of DFT, OPEC2 is much simpler than other automatic DFT synthesis systems. Another simplifying factor is that the rules for incorporating error checking circuitry into a design are well defined. This allows for an essentially algorithmic approach to the synthesis hence avoiding the need for a very complex expert system. A further benefit from this DFT technique is that the use of the embedded checkers, which in themselves are totally self-checking, eliminates the need for storing and comparing a large number of circuit responses for faulty and fault free circuits during the testing stage.

When using concurrent error detection techniques, the checkers are normally placed on the output of each function block in the circuit; however, this incurs an unnecessary overhead in area. The objective of OPEC2 is to limit the placement of checkers in the circuit to strategic positions so that the area overheads are reduced whilst ensuring that the fault detection properties of the scheme are not compromised.

Since OPEC2 is relatively simple it will be described in slightly more detail than the previous systems.

The basic criteria for the placement of error checkers in coded datapaths were summarized in Sellers (Sellers *et al*, 1968), and are as follows:-

1) The checkers should minimize undetected errors and faults.

2) Checkers should locate an error before it corrupts further data.

3) Checkers should improve fault resolution

4) Minimize the number of checkers used, hence reduce costs.

There are obviously trade-offs between satisfying the first three criteria and the last. However, a more systematic approach to checker placement was proposed by Wakerly (Wakerly, 1978). In this approach the circuit is considered to comprise a network of self-checking functional blocks. The criteria for checker placement can then be reduced to the single objective of placing checkers so that the circuit, overall, is self-checking.

The error checker placement rules implemented in OPEC2 are outlined below:

1) All 0-error detection lossless blocks must have checkers placed on all inputs. (Error detection lossless definition of a block specifies the number of erroneous inputs which can be applied without producing an undetectable error at the output)

2) If a datapath reconverges at n inputs of a t-error detection lossless block, checkers must cut at least (n-t) of the reconvergent fanout paths.

3) Every feedback loop in the datapath must be cut by at least one checker.

It is assumed that the function blocks are categorized as follows :-

1) Logic blocks are 0-error detection lossless.

2) Arithmetic blocks are 1-error detection lossless.

3) Multiplexers are n-error detection lossless.

4) Storage blocks are pseudo n-error detection lossless.

By placing checkers in accordance with the above rules, no non-code outputs can propagate in the circuit to create an incorrect code word at an output without an error being flagged by a checker.

OPEC2 is a rule based system implemented in Prolog. Its operation can be thought of as proceeding in two stages, the first being the acquisition of information about the circuit, and the second stage the placement of the error checkers. To start with, OPEC2 must be given a connectivity list for the circuit in question and a block function group specification. The connectivity list simply denotes which blocks are interconnected. The function group specification defines a block as belonging to any of the major functional block types (logic, arithmetic, etc). Two dummy function types are also included, input and output, which denote primary circuit input and output lines.

The information acquisition stage is concerned with the topology or connectivity of the circuit and is itself composed of two stages. Each individual stage extracts all the relevant information about the circuit connectivity and stores it in the circuit information database in a more usable form for later use. The first task is to identify all distinct feedback loops in the circuit. These are detected by flagging all circular pathways in the circuit. Then all the separate groups of reconvergent paths (multiple paths between any two points in the circuit) are detected. These are then stored, after a filtering operation which ensures that there are no redundant elements in the list of reconvergent paths. Once this stage is complete, OPEC2 proceeds with the placement of error checkers.

The placement of error checkers first starts with the placement of all compulsory checkers, these being the checkers on the inputs to logic blocks (rule 1). If any of these checkers cuts a feedback or reconvergent fanout path, the path in question is marked as being checked. The next stage in checker placement is the positioning of checkers to cut reconvergent fanout paths (rule 2). During this placement, if a choice of position is possible, priority is given to checkers which would also break a feedback loop. As before, after each checker is placed, all paths affected are marked as being checked. The last stage ensures that there are no unbroken feedback paths in the circuit. If this is not so, extra checkers are placed, with checkers being placed where they cut the largest number of loops.

A feature of OPEC2 is that it may also be used to verify a user placement of the error checkers. This feature is used by including in the circuit description file a list of placed checkers. OPEC2 makes use of this information, and will only place extra checkers if the user placement is not complete. This also allows the specification of a partial checker placement, if the designer decides that he needs a checker to be placed at some specific circuit location. An example of the input to the system and its output are shown below:-

```
%================================================
% Circuit definition   (for circuit in Figure 1.)
%================================================

% instance definitions: block_class( block_name ).

input( in1 ).
input( in2 ).
output( out ).
storage( b1 ).
storage( b2 ).
storage( b3 ).
storage( b4 ).
storage( b5 ).
arithmetic( b6 ).
multiplexer( b7 ).
logic(b8).

% connections: connect(block_name, block_name, net_number).

connect( in1, b1, 1).
connect( in2, b4, 2).
connect( b1, b2, 3).
connect( b1, b3, 3).
```

connect(b2, b5, 4).
connect(b2, b6, 4).
connect(b3, b6, 5).
connect(b4, b7, 6).
connect(b5, b8, 7).
connect(b6, b7, 8).
connect(b7, b8, 9).
connect(b8, out, 10).

The output generated by OPEC2 is as follows:-

Checker placed on output of block b5, net 7
Checker placed on output of block b7, net 9
Checker placed on output of block b3, net 5

Normally a checker would be placed in each net; for this circuit 10 checkers would thus be required. The OPEC2 placement of 3 checkers reduces the number of checkers by 67%.

With reference to Figure 1, the first two checkers, placed on nets 7 and 9 are due to the first rule: place checkers on all inputs to logic functions. The circuit contains three reconvergent fanout paths, block 1 to block 6, block 2 to block 8 and block 1 to block 8. The checkers placed by the first rule also break the reconvergent paths terminating on block 8, thus only leaving the reconvergent path between blocks 1 and 6 to be checked. Block 6 is an arithmetic block, which thus implies that it is 1-error detection lossless. There are two reconvergent paths to this function (both input lines), and thus one checker is required. Since there are no more constraints, the program arbitrarily decides to place the checker on net 5. With the placement of this checker, the circuit is totally self-checking and thus no more checkers are required.

OPEC2 also was applied to a modified version of a commercially available processor which normally require 40 checkers in the circuit, that is one on the output of each of the main function blocks. However, after being processed through OPEC2 the number of checkers required was 10, a reduction of 75%.

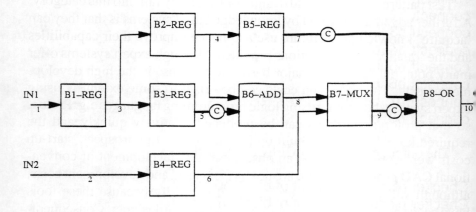

Figure 1   Circuit with checkers placed

5.5 Conclusions

The application of knowledge based systems to the design of VLSI circuits will have the greatest impact in assisting the engineer with aspects of the design which are not amenable to automation using algorithmic techniques. Typically these are areas(Preas and Lorenzetti, 1988) in which ,

1) A large problem solution space exists, which must be searched efficiently.

2) The problem to be solved has a number of conflicting constraints.

3) The knowledge of how to solve the problem is not completely understood.

4) There are a number of possible solutions to the problem , but no overall quantifying metric to decide which is the best.

Test pattern generation and design for testability fall into this category. A further advantage offered by knowledge based systems is that they can 'acquire' knowledge for future use and hence improve their capabilities and the 'quality' of the solutions provided.Although expert systems offer many potential benefits a major barrier to their use is the high developments costs. These start-up costs can be offset to some extent by using expert system shells or by using logic programming languages e.g. Prolog, so that prototype systems can be produced relatively quickly and the required knowledge bases built up incrementally. In retrospect start-up costs were considered to be an obstacle to the development of conventional CAD programs. These are now widely used and development costs, although still considerable, are accepted primarily because these tools have been shown to provide tangible benefits to the designer. Consequently, if development costs of expert systems are to be accepted these tools must demonstrate that they can also provided tangible benefits to the designer

5.6 References

ABADIR M.S., BREUER M.A., "A Knowledge Based System for Designing Testable VLSI Circuits", IEEE Design adn Test of Computers, Volume 2 Number 4, August 1984, pp56-68.

ABADIR M.S., "TIGER: Testability Insertion Guidance Expert System ", Digest of Papers, International Confernce on Computer Aided Design, CH2805, 1989 pp562-565.

ANGWIN G. et al, "The Need for Real Time Intelligence when Testing VLSI", Digest of Papers, International Test Conference, October 1984, pp752-9.

APFELBAUM L. "Improving In - Circuit Diagnosis of Analog Networks with Expert System Techniques", Digest of Papers, International Test Conference, 1986, pp947-53.

BELLON C. et al "An Intelligent Assistant for Test Program Generation: The SUPERCAT System", Digest of Papers, International Conference on Computer Aided Design, 1983, pp32-3.

BENDING M.J., "HITEST: A Knowledge Based Test Generation System", IEEE Design and Test of Computers, Volume 1 Number 2, May, 1984, pp83-92.

BOUNANNO G., SCIUTO D., " A Tool for Testability Analysis and Design for Testability of VLSI Devices", Proceedings of IFIP Workshop on Knowledge Based Expert Systems for Test and Diagnosis, September, 1988, pp76-79.

BREUER M.A., ZHU X., " A Knowledge Based System for Selecting a Test Methodology for a PLA", Proceeding 28th Design Automation Conference, June 1985, pp259-265.

CLOCKSIN W.F. and MELLISH C.S., "Programming in Prolog", Springer-Verlag, 1981.

DEVES P. et al, "Dedale: An Expert System for Troubleshooting Analogue Circuits", Digest of Papers, International Test Conference, 1987, pp596-94.

EICHELBER+GER E.B., LINDBLOOM E.,"Trends in VLSI Testing", VLSI, 1983, Elsevier Science Publications.

FUNG H.S. et al, "Design for Testability in a Silicon Compilation Environment", 22nd Design Automation Conference Proceedings, June 1985, pp190-6.

FUNG H.S. and HIRSCHHORN S., "An Automatic DFT System for the SILC Silicon Compiler", IEEE Design and Test of Computers, Volume 3 Number 1, February 1986, pp45-57.

GEADA J.M.C., RUSSELL G., "The Automatic Synthesis of Concurrent Self-Test Hardware for Safety-Critical VLSI Circuits", Journal of Semicustom ICs, Elsevier, March, 1991.

HARTLEY R.T. "CRIB: Computer FAult - Finding Through Knowledge Engineering", Computer, July 1986, pp68-76.

HAYES-ROTH F., "The Knowledge Based Expert System: A Tutorial", Computer, Volume 17 Number 9, September, 1984, pp11-28.

HILL F.J., HUEY B., "A Design Language Based Approach to Test Sequence Generation", Computer, Volume 10 Number 6, June 1977, pp 28-33.

PARRY P.S., *et al*, "An Application of Artificial Intelligence Techniques to VLSI Test", Proceedings Silicon Design Conference, July 1986, pp325-9.

PIPITONE F., "The FIS Electronic Troubleshooting System", Computer, July 1986, pp68-76.

PREAS B., LORENZETTI M. (Editors), "Physical Design Automation of VLSI Systems", Benjamin/Cummings Publishing Co., 1988.

RAMAMOORTHY C.V. et al, "Software Development Support for AI Programs", Computer, Volume 20 Number 1, January 1987, pp30-40.

ROBACH C. et al, "CATA: Computer Aided Test Analysis System", IEEE Design and Test of Computers, Volume 1 Number 2, May 1984, pp68-79.

ROBINSON G.D., "HITEST - Intelligent Test Generation", Digest of Papers, International Test Conference, October 1983, pp 311-23.

ROBINSON G.D., "Artificial Intelligence in Testing", Digest of Papers, International Test Conference, October 1984, pp200-203.

SAMAD M. A., FORTES J.A.B., "Explanation Capabilities in DEFT - A Design for Testability Expert System", Digest of Papers, International Test Conference, October 1986, pp954-963.

SELLERS F.F. et al, "Error Detecting Logic for Digital Computers", McGraw-Hill, 1968.

WAKERLY J., "Error Detecting Codes, Self-Checking Circuits and Ap0plications", Computer Design and Architecture Series, North-Holland, 1978.

WATERMAN D.A., "A Guide to Expert Systems", Addison-Wesley 1986.

WILKINSON A.J., "MIND: An Inside Look at an Expert System for Electronic Diagnosis", IEEE Design and Test of Computers, Volume 2 Number 4, August 1985, pp 69-77.

WILSON B.R., HNATEK E.R., "Problems Encountered in Developing VLSI Test Programming for COT", Digest of Papers, International Test Conference, October, 1984, PP778-778

Chapter 6
Knowledge Based Test Strategy Planning

C. Dislis, I. D. Dear and A. P. Ambler

6.1 *Introduction*

Over the past decade, integrated circuits have greatly increased in size and complexity. This increased complexity was not followed by a proportional increase in the number of pins, and as a result testing has become a time consuming and expensive operation. Testing costs can now form a significant part of the product's final cost. In addition, the quality and reliability of systems are serious issues, particularly for critical applications, and form an important part of a company's marketing policy.

In order to minimise testing costs as well as improve the quality and reliability of systems, testability requirements need to be viewed as part of the design specification, rather than as an issue to be considered at a later stage. In the case of electronic systems, testability should start in the design of the integrated circuits used. Early testability provision prevents escalating costs at later stages, as it is more cost effective to detect faults as early in a product's lifecycle as possible. Late detection of faulty devices, when they have already been incorporated into a system, involves the cost of processing a faulty device, as well as the cost of diagnosis and repair, and has a detrimental effect on the product's quality (Davies, 1982). However, testability provision does not come free and its costs need to be balanced against the resultant savings in testing cost. The rest of this chapter will examine ways of arriving at an optimum level of testability provision for integrated circuits, in a way that minimises its costs, while fulfilling the technical specification.

6.2 *Test Strategy Planning*

The problem of integrated circuit test needs to be viewed as part of the overall specification and design of the product. Design for testability

(DFT) methods are well documented and in many cases guarantee that the device will be fully testable. A fixed test strategy can be used, where a high level of testability is guaranteed if a set of design rules are followed, as in level sensitive scan design. This approach results in testable circuits, but can be inflexible. Another approach is to design the circuit following general testability guidelines which aim to improve the controllability and observability of nodes. This can be achieved by making sure that control lines are accessible, breaking long feedback paths and providing extra test points where necessary (Bennets, 1984). Such ad hoc methods make testing easier, but depend to a large extend on the designer's perception of what constitutes good DFT practice and do not guarantee testable chips.

It is becoming apparent that hybrid DFT methods may be more effective, both in cost and testability terms, than either ad hoc or fixed strategies, especially as DFT methods are often optimised for specific design classes, such as PLAs or RAMs. The implementation of a hybrid strategy, however, depends on the designer's expertise and detailed knowledge of test methods.

In this chapter the term test strategy planning is used to describe the choices of design for test methods for a specific application. The problem to be solved is a complex one, as a large variety of parameters need to be taken into account, such as the design specification, test access, test pattern generation systems and the implementation details of design for test methods.

The problem can be tackled at several levels. One method is to optimise the fixed strategy approach. For example, a more efficient use of the scan method is to target only the flip flops which are critical in testability terms. The choice can be based on a set of heuristics. For example, a set of flip flops which break the cyclic structure of the circuit can be chosen. It is the cyclic structure of the circuit, rather than simply its sequential depth that create testability problems (Cheng and Agrawal, 1990). Alternatively the choice can be based on the difficulty of generating test patterns for specific nodes. The Intelligen test pattern generator (Racal-Redac) allows the user to place scan type cells at nodes which have created problems in test pattern generation. This is an example of dynamic testability analysis, but static testability analysis methods can also be used to select a flip flop set.

Although these methods offer an improvement on the full scan strategy, they still suffer from the same drawbacks of single strategy test. The alternative is a test strategy planning system that allows the user to choose from a variety of test strategies. This can be done by allocating test resources (LFSRs, MISRs etc) to critical parts of the design, or by allocating test methods to functional blocks. It is multiple strategy systems and their use of knowledge based methods that are of particular interest in this chapter.

6.3 Knowledge Based Systems and Test Planning

It is already apparent that the test strategy planning problem is a complex one due to the variety of issues that need to be considered. It is essentially the task of a testability expert. It is necessary to have knowledge of the testability methods that are available, and of the ways these methods can be incorporated into an existing (or ongoing) design. Looking at the problem from this point of view, the value of knowledge based systems becomes clear. A test strategy planning system can consist of a knowledge base of design for test methods together with a set of rules which govern test methods application, and some means of evaluation of test strategies. The following sections will examine some existing systems (Breuer et al, 1989) and will describe current work in this area.

6.3.1 Current systems

There are several knowledge based systems which perform test strategy planning for integrated circuits to various degrees. Possibly the best known one is TDES (Abadir and Breuer, 1985), developed at the University of Southern California. A later system, called TIGER (Abadir, 1989) is a continuation of the TDES work. In TDES/TIGER, a register transfer level design is partitioned by the system into testable units called kernels. The knowledge base consists of testable design methodologies (TDMs) which are stored in a frame-based representation, each TDM being described by a set of attribute-value pairs. The knowledge stored about each TDM includes the test structures necessary to implements it, a parameterised test plan, and the effects on the circuit in terms of area overhead and test time. The system evaluates testing possibilities by using different sets of driver/receiver registers by utilising I-paths, which are transparent paths for information transfer during test. The system evaluates possible configurations (or embeddings) of TDMs for each kernel, and makes a cost/benefit analysis, based on the structure of the TDM, the circuit structure, and a set of user specified parameter weights. A set of scores is produced, based on which an optimal test strategy is chosen. Along the same lines, PLA-TSS (Zhu and Breuer, 1987) is an expert system for testable PLA design. In this case TDM attributes are a function of PLA size and test requirements. The system attempts to find the test methods which are closest to the user's constraints and requirements by using penalty/credit functions for each attribute and combining these into a score in a linear fashion. Backtracking is possible when the system fails due to over-constrained requirements, in which case an analysis is performed to identify the most critical reasons for failure.

The Alvey BIST system (Jones and Baker, 1987) was a test strategy planning system programmed in LOOPS, a Lisp based language. The system attempted to use test resources in the circuit such as shift registers and BILBOs. These may exist already or can be added on later. Each test resource is given a weighting depending on the number of blocks that can use it, and test resources with highest weighting are used first. Constraints of area overhead and test time are also considered, and backtracking is supported. The knowledge was derived from a group of testability experts.

Design for testability methods can be made part of the design synthesis route, and this is what the Silc automatic design for testability system aims for. The system, named TESTPERT (Fung and Hirschhorn, 1986), links design and testability issues in the silicon compilation process. Testability 'bottlenecks' are recognised by a testability evaluator, which works within user defined bounds of area overhead and test time. Special testability structures are available, together with cost/benefit attributes. The system works by manipulating finite state machine descriptions, and testable structures are added as early in the design process as possible, based on the benefit attributes and user constraints.

6.3.2 Test strategy description and evaluation

The systems mentioned in the previous section all take different approaches to test strategy planning. The representation of test method knowledge varies accordingly, but in most cases a frame based representation is used. Another common feature is that the user is required to aid the evaluation process by supplying relative weights for parameters such as area overhead and test time. However, there are other aspects which may affect the choice of test strategy, such as the extra development effort that will possibly be needed, the test pattern generation effort saved, and the availability of automatic test equipment. Also, by asking the user for a weighting system, the systems assumes that the information is objective. This may not always be the case, as the priorities of a test engineer are not necessarily the same as those of a designer. The former may ask the system to find the best strategy that minimises test time while the latter may opt for a minimisation of logic overhead. As a result, the answers are likely to be different.

In some systems the scores of strategies are combined using a cost function in order to be able to compare parameters which are not similar. However, it would be better to find an objective common reference point so that dissimilar parameters can be compared directly. As most companies evaluate projects in terms of financial costs and savings, it seems that if parameters can be translated into financial terms they will be directly comparable. The research taking place at Brunel University uses cost modelling as a method of test strategy evaluation, and is described in more detail in the following sections.

6.4 Cost Modelling and Test Strategy Evaluation

The Brunel test strategy planning system, called BEAST (Brunel Economic Aids for Strategic Test) was developed under the Alvey CAD042 program, It is targeted at the standard cell / macro cell design approach, as it is then easier to obtain information directly from the CAD system in the early stages of the design, and test methods are often optimised for specific cell types. The architecture of BEAST is shown in

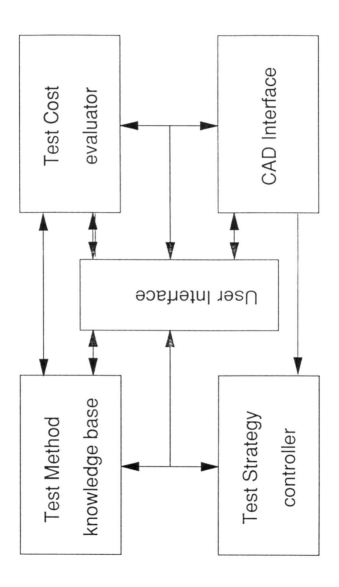

Figure 1 Beast architecture

Figure 1. The knowledge base holds information on the requirements and implications of test methods for specific cell types. This, together with the design description is used by the test strategy controller in the formulation of test strategies. The test cost evaluator is the cost model used to compare test strategies based on their respective costs. The individual 'parts have been kept largely independent of each other, so that they can be accessed separately.

6.4.1 The test method knowledge base

The knowledge base of test methods holds data on test methods applicable to different functional blocks (Dear et al, 1989). It contains a list of test dependent parameters which are passed to the cost model for the evaluation of different strategies, as well as a list of additional test structures required for the implementation of the chosen test method. Specifically, the parameters stored include a list of test structures needed to implement the chosen strategy, such as extra latches or control lines. The area overhead of the method, its logic (gate) and associated routing overhead are also included, as well as any overhead in input and output lines which may affect the number of pins in the final circuit. Parameters are stored as attribute-value pairs, although in some cases an equation needs to be evaluated before a parameter value can be determined. This can happen in parameters which are both design and test dependent such as functional and routing area overhead. In this case, area overhead is calculated using a parameterised floorplan of each functional block (Miles et al, 1988).

Other test related parameters are involved in modelling test pattern generation effort, and details about preferred test pattern generation methods or algorithms are also held in the knowledge base. A flag identifies self test methods. In the test application area, specific automatic test equipment requirements are listed, such as speed of test vector application or minimum memory size per pin. Finally, the maximum fault cover achievable by the method is included. The knowledge base also contains information that is not passed directly to the cost model, but is of interest to the designer for implementation, or to the test engineer when formulating a test plan, such as whether the number of test patterns is function dependent. The system acts as an advisor, and does not in itself implement the preferred test strategy into the design description.

The design information is also held in a knowledge base, and the relevant data is provided by the user at the start of the test planning session. The design knowledge base contains information on the numbers and types of functional blocks used, as well as the connectivity of blocks, and information about their accessibility for test purposes.

6.4.2 Test strategy evaluation

The economics model which is used for test strategy evaluation, is a parameterised cost model, structured hierarchically in order to allow evaluation of test strategies from the early stages of the design process (Dislis *et al*, 1989). At the highest level of hierarchy, costs throughout the component, board and system stages are calculated, as well as field and maintenance costs. At this level, broad marketing and test strategies can be examined. It may be found, for example, that the increased reliability that results from extra expenditure in the component test stage may drastically reduce field maintenance costs and therefore be a justifiable expense. At lower levels, when more information is available on the partitioned design, a more detailed approach is necessary in order to make specific test strategy decisions. Therefore costs of design, manufacture and test for each partitioned are considered. This is the level used for most of the test strategy planning process for integrated circuits in this work.

The main purpose of the economics model is to compare test strategies, rather than make absolute cost predictions. This approach tends to minimise errors due to variations in the input parameters. The data used by the model are specific to the user company, and the accuracy of the model predictions would obviously depend on the accuracy of the input values. Therefore, the confidence in the absolute model predictions should increase with use.

The component level model calculates design, production and test costs for a partitioned design, with the ability to predict costs associated with each partition. This approach agrees with the standard cell oriented structure of the test planner. The structure and main parameters of the model are summarised in Tables 1 to 3.

Design costs are modelled as a function of circuit size, number of different functions of the device, and the originality of the design, which considers whether a similar design has been implemented in the past. Labour rates and equipment costs also influence the total design cost.

Production costs take into account mask or e-beam equipment cost, as well as throughput, process yield, manufacturing equipment costs, packaging and manpower costs. Packaging costs are related to the type of package and the number of pins required. If there are unused pins in the package used, then these can be utilised for test purposes for no extra cost. However, extra test pins may mean that another package size is required. The model contains a database of package types, so that the cost of the required package can be accessed automatically. A variety of yield models, either theoretical or empirical, can be used to model the production process.

Design Costs	
Design Parameters for I.C.	Functions Originality
Design Parameters per Partition	Gate Count Gate Overhead Area and Routing Area Area and Routing Overhead Originality
Labour and Equipment	Number of Designers Labour Rates Equipment Rates

Table 1 Design cost modelling

Production Cost	
Yield Modelling	Defect Density Yield Modelling Early Life Yield Number of Devices Required
Mask Costs	Master Mask Cost Working Mask Cost Number of Masks
Silicon Cost	Wafer Cost Wafer Area
Labour Cost Equipment Cost	Labour Rates Manpower Equipment Rates
Packaging Cost	Number of Pins Package Type/Cost Database

Table 2 Production cost modelling

Test Cost	
Test Pattern Generation Per Partition	Self Test? Chip Complexibility Fault Cover Accessability
Test Application	Hourly ATE Cost Throughput Test Programming Labour Rates

Table 3 Test cost modelling

Test costs are treated in two parts, test generation and test application. Test generation costs are a function of circuit type and structure as well as the position of a sub-circuit in the design, ie, its accessibility for test. For instance, test generation for a RAM block can be entirely algorithmic, and test costs are mainly due to test application because of relatively large test length. However, if the RAM is not easily accessible, extra costs may be involved in propagating the test vectors to it. Test application costs are linked to ATE running costs, test programming time, test application time, manpower costs, as well as test application strategies such as batch sizes and time to process each batch.

6.5 Test Method Selection

A set of designs were examined in order to analyse the economic effect of the test strategy used, the type of functional block used in the design, and the effect of overall chip size and production volume on global cost. Two types of functional block were considered, a PLA and a RAM, both occupying approximately the same area. The test methods employed were: no design for test, full scan and self test, using BILBOs for the PLA and the Illman method (Illman, 1986) for the RAM. The area overheads were calculated from a parameterised floorplan, and are implementation dependent. The types, size and area overhead for each block are summarised in Table 4. PLA sizes are given in terms of number of inputs, outputs and product terms. In order to assess the effect of the overall chip size on functional block costs, each block was embedded in both a 0.25 cm^2 and a $1 cm^2$ chip. Cost predictions refer to the block, rather than the whole chip. However, because yield is dependent on chip area, the overall chip size will affect functional block costs.

Type	Size	Area (cm^2)	Overhead Scan (%)	Overhead Self-test (%)
PLA	10, 10, 10 10, 10, 80	0.001 0.005	26 9	66 25
RAM	64 bits 256 bits	0.001 0.004	20 6	35 14

Table 4 Functional block information

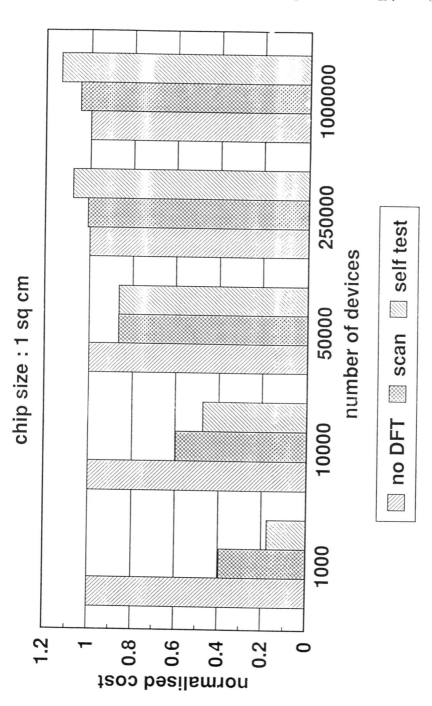

Figure 2 PLA test method comparison

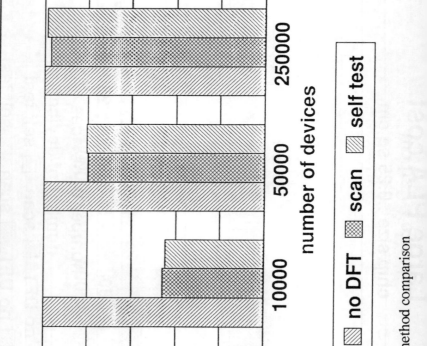

Figure 3 RAM test method comparison

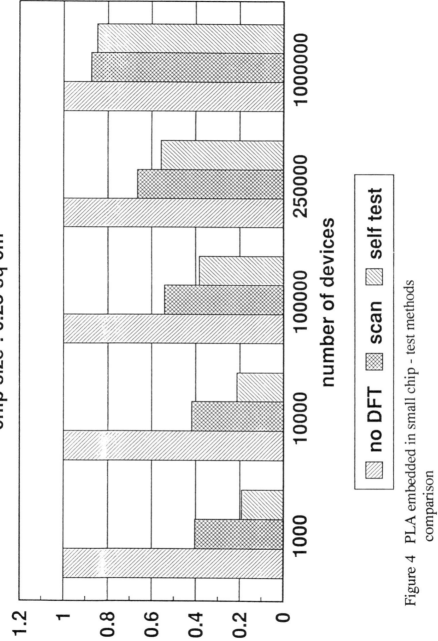

Figure 4 PLA embedded in small chip - test methods comparison

136 Knowledge based test strategy planning

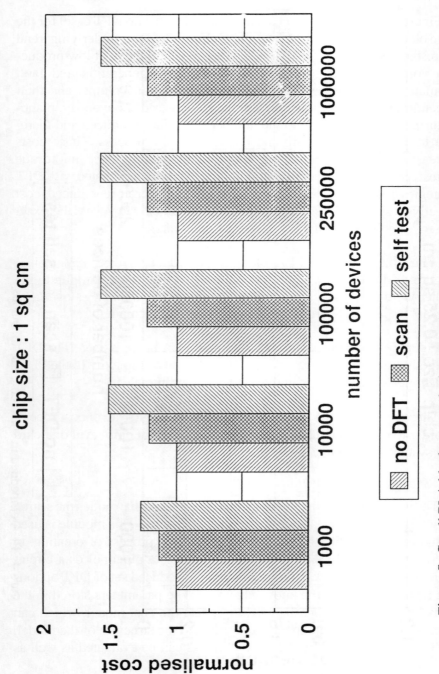

Figure 5 Small PLA block test methods comparison

Figures 2 and 3 show the costs (normalised to the no DFT case) for the large blocks embedded in the larger size chips. The same underlying trend is present, namely that DFT strategies are most attractive at low production volumes. Fixed costs such as design, test generation and fault simulation are amortised over the entire production volume, and their effect diminishes as production volume is increased. Conversely, manufacturing costs are related to the number of devices produced and therefore to the area of the chip and the yield of the process. These costs increase with production volume and, for large volumes, account for the largest fraction of the overall cost. The area overhead associated with DFT strategies results in a reduction in yield and therefore reduced cost effectiveness of DFT strategies for large volumes. However, the exact crossover point will vary.

When the same size PLA block is embedded in the smaller chip, (Figure 4), DFT is now a more attractive option throughout, due to the higher yield for small chips. The higher yield means that area overhead has less impact on overall cost, while test savings remain.

Figure 5 shows the costs when the small PLA block is used. The DFT methods are now more expensive, due to the low complexity of the PLA, which means that savings in the test area are minimal.

These results can be further evaluated in the higher level cost model, in order to determine the effect of decisions taken at chip level on board and system level costs.

6.6 <u>The Test Strategy Controller</u>

The test strategy controller is an inference engine that guides the search for an optimal test plan according to a set of criteria. This module is used in the initial selection of test strategies and the process of creating an overall test plan. Once an architecture (such as a pipeline) or a circuit structure such as a PLA have been targeted, a reduced set of DFT options can be retrieved form the knowledge base. The parameters stored in the knowledge base can then be used to update the cost model, which can then be used to evaluate the methods. Due to the structure of the model, DFT methods for specific functional blocks can be evaluated as well as global strategies. One of the functions of the test strategy controller is to relate the DFT methods selected for functional blocks in isolation and produce a globally optimised test strategy plan for the device.

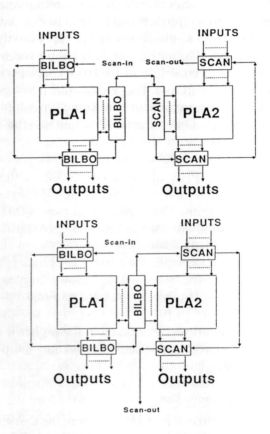

Figure 6 Test strategy optimisation example

The optimisation of the final test strategy is currently treated as a post process to the strategy selection. The test strategy optimisation involves the analysis of the final test strategy in order to determine if certain circuit structures can be used to test more than one block. For example, a BILBO register can also act as a scan register to test a neighbouring block. An example of the optimisation process can be seen in Figure 6. The optimisation uses a set of heuristics, which rely on the connectivity of the blocks, as well as their transfer functions. Transparency for test purposes is also considered. It is likely that different heuristics will perform better for specific architectures. Therefore, the selection of heuristics used can be related to the high level architectural information supplied during the early stages of the design. Other aspects that need to be taken into account in the optimisation process are the test requirements, size and functionality of blocks. The cost model can again be used to evaluate different optimisation scenarios.

Currently, the selection of test methods is entirely cost based. However, it is relatively simple to base the selection on other factors such as fault cover, with cost being a secondary consideration. This may be the case for critical applications.

Once test methods are selected and ordered according to the chosen criteria, a full test plan can be formed. With a small design and a short list of test methods, test plans can be evaluated exhaustively. However, this rapidly becomes an impractical solution as designs get larger. Different methods can be employed to reduce the search space. This process can start with the initial test method selection, immediately discounting methods which are too expensive. One way to reduce the test method set, for example, is to use say the three most cost effective methods for each block. One of the problems encountered in test strategy planning is that of local minima, and how they can be recognised as such. The order that blocks are evaluated is significant, so strategies which target the largest/least testable/least accessible block can be employed. Another method is to assess the connectivity complexity of each block, in order to find critical blocks in the design, ie the ones that control the largest number of paths, and target them first. The aim is to keep backtracking to a minimum, in order to minimise the time taken to arrive at an optimal plan.

6.7 Conclusions

This chapter has described some current knowledge based systems to aid design for testability, and described research in this area taking place at Brunel University. It has also attempted to highlight specific problem areas in test strategy planning, and presented an economics based approach to the problem of test strategy evaluation. The system described here was constructed as a research tool, but the concepts and test strategy planning methods were used in a current ESPRIT project for gate array test strategy planning, which is now under industrial evaluation (Dislis *et al*, 1991). The basis of the test strategy planner is the cost modelling employed, which the authors believe is a reliable method for making realistic comparisons of design for test methodologies. The results of the economics based planning have so far agreed with test experts' approaches.

Test strategy planning is a complex problem, and it is now becoming urgent to arrive at an automated method for an optimal solution. Knowledge based systems are well placed to handle this problem, and if realistic methods for strategy evaluation are employed, a full, industrial test strategy planner should soon become a reality.

6.8 *References*

ABADIR M.S., BREUER M.A., "A Knowledge Based System for Designing Testable VLSI Chips', IEEE Design & Test of Computers, Vol. 2, No 4, August 1985, pp 56-68.

ABADIR M.,"TIGER: Testability Insertion Guidance Expert System", Proc. IEEE International Conference on Computer Aided Design, 1989.

BENNETTS R.G., "Design of Testable Logic Circuits", Addison-Wesley Publishing Company, 1984.

BREUER M.A., GUPTA RAJESH, GUPTA RAJIV, LEE K.J., LIEN J.C., "Knowledge Based Systems for Test and Diagnosis", Knowledge Based Systems for Test and Diagnosis, North Holland, 1989.

CHENG K.T., AGRAWAL V.,"A Partial Scan Method for Sequential Circuits with Feedback", IEEE Transactions on Computers, vol 39, no 4, April 1990, pp 544-548.

DAVIES B., "The Economics of Automatic Testing", Mc-Graw Hill, 1982.

DEAR I.D., DISLIS C., LAU S.C., MILES J.R., AMBLER A.P., "Hierarchical Testability Measurement and Design for Test Selection by Cost Prediction". Proc. European Test Conference 1988.

DISLIS C., DEAR I.D., MILES J.R., LAU S.C., AMBLER A.P., "Cost Analysis of Test Method Environments", Proc IEEE International Test Conference, 1989, pp 875-883.

DISLIS C., DICK J., AMBLER A.P., "An Economics Base Test Strategy Planner for VLSI Design", Proc. European Test Conference, 1991.

FUNG H.S., HIRSCHHORN S., "An Automatic DFT System for the Silc Silicon Compiler", IEEE Design and Test of Computers, vol 3, no 1, February 1986, pp 45-57.

ILLMAN R.J, "Design of a Self Testing RAM", Proc. Silicon Design Conference, 1986, pp 439-446.

JONES N.A., BAKER K., "A Knowledge Based System Tool for High Level BIST Design", Microprocessors and Microsystems, vol 11, no 1, January/February 1987, pp 35-40.

MILES J.R., AMBLER A.P., TOTTON K.A.E., "Estimation of Area and Performance Overheads for Testable VLSI Circuits", Proc IEEE International Conference on Computer Design, 1988, pp 402-407.

ZHU X., BREUER M.A., "A knowledge Based System for Selecting a Test Methodology for a PLA", Proc. 22nd Design Automation Conference, June 1985, pp 259-265.

Chapter 7
HIT: A Hierarchical Integrated Test Methodology

C. A. Njinda and W. R. Moore

7.1 *Introduction*

As increasing use is made of silicon compilers and circuits modelled using VHDL or EDIF, there would appear to be a growing need to produce test generation tools that function in a similar way, hiding most of the information about the lower levels of the design hierarchy. For such tools to be cost effective, it is necessary that they provide enough information so that the target chip is adequately tested and that the test time is kept short. These are two conflicting requirements, because to obtain sufficient fault coverage we need detailed information about the circuit while to reduce the test generation time, we need to decrease the amount of information to be considered so as to reduce complexity. It is apparent that the latter can only be achieved if designs are modelled at higher levels using functional elements such as ALUs, RAMs, PLAs, systolic arrays, registers, word-level logic gates etc..

Various approaches have been developed to generate test vectors or apply built-in self-test (BIST) techniques to circuits using high-level circuit information. Melgura (Melgura, 1987), Shen (Shen and Su, 1984) and others (Villar and Bracho, 1987 and Thatte and Abraham, 1978) have based their methods on the use of functional fault models. Unfortunately, though elegant, such methods are prone to produce an overly large set of test vectors and/or a poor coverage of the real physical faults. For example, consider a hypothetical circuit composed of two registers A and B connected to a processor. The functional level faults associated with the transfer of data from A to the processor can be modelled as:

1) the data in B instead of A is transferred to the processor, or

1) both the data of A and B are transferred to the processor at the same time, or

2) the data in A cannot be transferred to the processor.

It is clear that this kind of fault hypothesis will not account for the real physical faults such as stuck-at, stuck-open or bridging faults in the A register. This results in a poor fault coverage. Apart from the difficulty of modelling faults, it is very difficult to explicitly obtain the behavioural description of every circuit and this limits the applicability of these approaches. If we consider the case of an n-bit serial adder, for it to be tested functionally all 2^{2n} vectors must be applied to the inputs. However, it can be shown that to cover all stuck-at faults in such an adder of any length only seven test vectors are required.

Other approaches to hierarchical testing implement designs from testable macro cells (Lenstra and Spaanenburg, 1989). Each macro is associated with a test controller that apart from controlling the functionality of the macro cell can also be used to direct the test process. Test structures are added, if needed, to increase the system's testability by improving access to the macros' inputs and outputs. The main difficulty in using testable macros when designing circuits is caused by the fact that there are many different test structures for each kind of cell. This increases the knowledge pool to be handled and hence the complexity. Furthermore, there is the possibility of sharing test structures between different cells. If sharing is not considered, the solution obtained might not be the best. To be able to test all macros some interface elements may also be required. Because no global approach is available to select these interface elements, it is possible that the area overhead will be very high.

Another approach to the hierarchical integrated test methodology was pioneered by the TDES system (Abadir and Breuer, 1985). In this system, a circuit is modelled at the register transfer level. Making a circuit testable is a process of selecting a set of registers in the design to be configured as Random Pattern Generators (RPG) and Signature Analyzers (SA) such that each combinational circuit is considered testable. A feasible test plan (test execution process) for the the design is also obtained. Other systems like TIGER, BIDES etc. (Kim *et al*, 1988 and Abadir, 1989), have been developed based on the TDES philosophy. Actually, none of these systems extend the philosophy of TDES but they do try to consider real world issues.

In trying to obtain the test vectors of a circuit, a designer would normally use some automatic test pattern generation (ATPG) tool. The result from such a tool is usually a set of test vectors, the list of faults covered and the list of undetected faults. The reasons why faults are not detected are usually not given. If the fault coverage is not adequate, one of the BIST tools (e.g TDES) might be used to make the design self-testable. This might result in a significant degradation of performance making the solution unacceptable to the designer. However, it is possible that if the reasons for failure when using the ATPG tool are considered, this degradation in performance might be avoided. Only certain portions of the design might need to be modified if this failure data can be made available to the designer.

The essence of HIT is therefore to provide a design environment overcoming the traditional separation of design and test development. The integration of tools such as simulation, testability analysis, automatic test pattern generation, DFT and BIST techniques within the same environment enables the designer to carry out these tasks himself. Hence the testability of the design is considered at an early stage of the design phase thus reducing the ATPG cost, enabling efficient application of scan and/or BIST techniques, and reducing the design cycle.

To achieve the above in HIT, we use a knowledge base to store the properties of previously characterized circuit elements called *cells*. Cells are divided into four categories: *combinational* (e.g. PLA, random logic), *memory* (e.g. RAM, CAM, ROM), *register* (e.g. latches, flip-flops) and *bus structures* (e.g. fan-outs, bus concatenation, sign extend, bus select). Each of the cells has various attributes attached such as test vectors or a test strategy, testability estimates (reflecting the ease of vector propagation or justification through the cell) and functionality. A structure consisting of these cells is termed a *design*. Once a design has been fully characterized (test process, testability estimates and functionality established) it can be used as a cell in a future design thus reflecting the hierarchy in HIT.

The test generation process in HIT relies on being able to establish a Sensitization path (*S-path*) from the selected cell (*target*) to the design's primary outputs and a Transmission path (*T-path*) from the design's primary inputs to the input(s) of the target. This operation is termed *path-planning*. For this process to be successful, cells are characterized to illustrate how data can be propagated through them unchanged (an *identity mode* or *i-mode*). The properties of i-modes have been discussed by Breuer (Abadir and Breuer, 1985 and Breuer et al, 1988). Once this process is complete, the set of cells that is considered testable is determined. A potential fault coverage figure is computed and a *rule-knowledge base* is consulted to help advise the designer of suitable modifications to improve the fault coverage and testability of the design. A similar graph approached to that used by Abadir (Abadir and Breuer, 1985) is used to help represent the constraints posed by various cells. HIT has the following potential advantages:

1) reduction in test effort and efficiency,

2) possibility of generating test patterns at arbitrary levels of hierarchy,

3) creation of a dialogue between the designer and testability expert which in most cases will result in an improved design.

In the following sections we present a brief overview of HIT as well as other related work on the subject.

7.2 System Overview

HIT is partitioned into four main blocks:

1) Cell characterization,

2) Design testability estimators,

3) Design partitioning, and

4) High level test pattern generation and design modification.

7.2.1 Cell characterization

Cell characterization is aimed at obtaining the functional description, the test vectors or test strategies and the testability estimates of each cell in the *cell knowledge base*. This information only needs to be computed when a new cell is to be stored in the cell knowledge base. If it is possible to obtain a functional description of a cell, the data obtained is used to represent the test vector propagation/justification modes. For example, the functional description of two cells is presented in Figure 1.

Since the designs that we have considered so far are mainly signal processing chips, it is usually easy to get the data in Figure 1. However, if a cell is designed using random logic, the above data cannot be easily specified. There are two situations to be considered here. Firstly, if the cell has a one-to-one mapping between the inputs and outputs, then any fault at the input will manifest itself at the output. In this situation we would still assume there exists an i-mode from the input to the output of that cell. In the second situation where a one-to-one mapping cannot be guaranteed, probability estimates are used to determine information transmission through the cell. We are also working on integrating lower level techniques to overcome this problem (Micalef and Moore, 1991). Bhattacharya and Hayes (Bhattacharya and Hayes, 1990 and Murray and Hayes, 1988) have presented other methods of representing the input/output of circuits with different levels of complexity.

Since the test generation and design modification strategy in HIT depends on the ability to compute suitable paths for vector propagation and justification, heuristics are required to direct the search for feasible solutions. In FAN (Fujiwara and Shimono, 1983) and other gate-level test generation algorithms (Abadir, 1989, Goldstein, 1987 and Roth, 1983), these heuristics are base on the 0-1 controllability and observability of individual bit lines. When cells are modelled at the register transfer level, this 0-1 controllability or observability of the bit lines in a bus conveys very little information. Also to get these values is very computationally intensive. Since we are only interested in approximate values to direct our search, we have adopted an approach that depends only on the complexity and i-mode data of each cell in the design. Instead of computing these values for single bits, there are determined for the entire bus. If a bus is a single bit, the values obtained will be similar to those for gate-level designs. The reader is referred to (Njinda and Moore, 1991 and Thearling and Abraham, 1989) for a more detailed discussion of design heuristics for circuits modelled at the register transfer level.

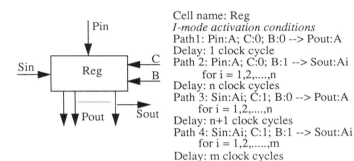

Cell name: Reg
I-mode activation conditions
Path1: Pin:A; C:0; B:0 --> Pout:A
Delay: 1 clock cycle
Path 2: Pin:A; C:0; B:1 --> Sout:Ai
for i = 1,2,.....,n
Delay: n clock cycles
Path 3: Sin:Ai; C:1; B:0 --> Pout:A
for i = 1,2,....,n
Delay: n+1 clock cycles
Path 4: Sin:Ai; C:1; B:1 --> Sout:Ai
for i = 1,2,.....,m
Delay: m clock cycles

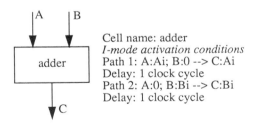

Cell name: adder
I-mode activation conditions
Path 1: A:Ai; B:0 --> C:Ai
Delay: 1 clock cycle
Path 2: A:0; B:Bi --> C:Bi
Delay: 1 clock cycle

Figure 1 Functional description of two cells

For each cell in the design, a *figure of merit* (ε) between 0 and 1 is attached to illustrate the cost of transferring data through that cell. If this value is close to 1, then it is very easy to transfer data unchanged through that cell. However, if ε is close to zero it is difficult to determine if data can be transferred through the cell unchanged. The computation of ε is the product of two parameters defined as the *transparency* (τ) and *sequentiality* (σ) of the cell. τ is used to represent the ratio of the actual number of vectors that can be obtained on the output of a cell to the maximum possible. So for an n-bit output where only m out of 2^n bit patterns can be obtained then τ equals $\dfrac{m}{2^n}$. For cells of different complexities, τ is not adequate to specify the propagation characteristics. For example, an adder and a shift register cell both have transparency values of 1. However, to propagate data through the adder only a single clock cycle is required while through the shift register n-clock cycles are needed (where n is the width of the register). This information is taken into account by the second parameter termed the sequentiality, σ, that is used to express the difficulty of transferring data through a cell.

ε is also used to express the testability value for a cell in isolation. More detailed discussion of the testability estimates for design are given in (Njinda and Moore, 1991, Thearling and Abraham, 1989, Bennetts, 1984 and Goldstein, 1987).

For each cell in the knowledge base, a set of test vectors or a test strategy is required. The fault model used in HIT is called a *cell fault model*. That is, for each cell in the design, the designer decides which fault model to use depending on the accuracy required, the complexity of the design at hand and the resources available for testing. Thus for regular arrays, such as systolic arrays, PLAs or RAMs, a test strategy to reflect their regularity and likely defects is selected (Marnane and Moore, 1989, Somenzi and Gai, 1986, Illman, 1986 and Salick *et al*, 1986). A new strategy can also be used provided that the test vectors are stored in a format that can be easily used by HIT. Test vectors are represented (stored) in a graph format called a *testing graph*. This is a directed graph relating the input vectors and the faults covered to the output vectors. Because of memory limitations, in most cases only the test generation strategy of cell is stored alongside the functionality and the testability information. The test strategy can be used in generating test vectors for the cell when required. This idea of being able to select different test strategies for different parts of the design provides a reasonable amount of flexibility to the designer, since he is able to use a test method based on the situation at hand.

7.2.2 Design testability estimators

In the last section we presented the testability estimators for a cell in isolation. When cells are combined together in a design, the way in which data can be propagated to different regions in the design is dependent on the the testability data of each cell in the design. Since in most cases we are required to control the busses in the design to specific values, we will model the testability of a design in terms of the controllability and observability of each bus in the design. A more detailed discussion of controllability and observability of busses when designs are modelled at the register transfer level is presented in (Njinda and Moore, 1991 and Thearling and Abraham, 1989). To be able to control the output of cell, we must be able to control the input values to the cell and transfer the necessary data to the output. Likewise to observe the value on the input bus to a cell, we must be able to control other input busses to the cell such that the data on the selected input is transferred to the output. Therefore the controllability of the output bus i of a cell is:

$$Con_i = \varepsilon \cdot \left(\frac{\sum_{j=1}^{no.\ of\ inputs} Con_j}{no.\ of\ inputs} \right)$$

Similarly the observability of the input bus to a cell is

$$Obs_j = \varepsilon \cdot \left(\frac{\sum_{j \neq i}^{no.\ of\ inputs} Con_j}{no.\ of\ inputs - 1} \right)$$

The use of these estimates will be explained in the next two sections.

7.2.3 Design partitioning

Partitioning in HIT is accomplished by using either *ad hoc* techniques or the controllability and observability estimates computed in the previous section. Partitioning using *ad hoc* techniques is driven by the following two rules:

1) *If* a feedback loop covers f or more cells, *then* use a multiplexer structure to break the loop, *else* group the cells surrounded by the loop into a single cell.

2) *If* the number of inputs to a combinational logic cell is greater than n (n fixed by the designer - typically 16), *then* divide the cell into two or more subcells with less than n inputs.

In the examples considered so far we have restricted the value of f to 1. In this case no feedback loops are allowed in design. This however increases the area overhead required for testing. If f is greater than 1, a sequential test pattern generator is required. This also increases the cost and difficulty of test generation. Furthermore, because of the feedback in the new cell, a sequence of vectors might be required to test some faults. This results in a larger memory requirement for test pattern storage. The number of multiplexer structures that can be added to a design depends on the extra area available for testing. We could model this partitioning approach by using a cost function (in terms of area overhead and test generation cost). Assume the cost of adding a multiplexer structure is x units, the cost of generating a test for a combinational cell is z units and for a sequential cell is y units. Then the objective function when using this partitioning approach would be to minimize both $\sum_{i} x_i$, where x_i is the area of multiplexer structure i, and $\sum_{j}^{N1} z_j + \sum_{k}^{N2} y_k$, where N1 is the number of combinational cells and N2 is the number of sequential cells

HIT: a hierarchical integrated test methodology 151

in the design. Note that for a bus-structure the test generation cost is zero. The second rule above is used to exploit the partitioning approach developed by Jone (Jone and Papachristou, 1989). In this approach a cell is broken up into subcells by using transparent latches such that the input dependency of each output bit of the cell is no more than n. n can be fixed by the designer. Other partitioning methods like sensitize partitioning (Urdell and McCluskey, 1987) and partitioning for exhaustive testing (Srinivisan et al, 1991) can also be used.

Our second partitioning approach is to use the controllability and observability estimates of the busses in the design. To test a cell in the design, we must be able to control the values on the input busses and observe the output busses. If the controllability of a bus is below a certain threshold, it will be difficult to propagate test vectors to a cell driven by that bus, while if the observability is below a fixed threshold, the data on the output of the cell driving that bus cannot be observed. The controllability case can be remedied by adding multiplexer structures while the observability situation is fixed by connecting the selected bus to an output pin. These two modifications will incur cost in terms of area overhead and extra input/output pins. So the objective function in this situation is to minimize $\sum_i x_i$ and the number of extra input/output pins.

7.2.4 High level test generation and design modification

The high level test generation and design modification (HTGDM) of HIT is the main work horse of the system. It does all the path-planning, symbolic computations and design modifications using various rules available in a *rule-knowledge base*. During the path-planning exercise, the cell i-modes, controllability and observability of the busses are used to determine the most suitable path for vector propagation and justification (Njinda and Moore, 1991). Once such paths have been established, they are analyzed symbolically to create a test program for the target. As stated earlier, because of the complexity of some cells in the design, it might not be possible to establish all the required paths. In such cases, DFT and BIST modifications are suggested and the possible cost of implementing any of the suggested techniques is computed. These suggestions are based on pre-programmed rules. By being able to guide the designer through various DFT and BIST techniques, he/she is in a better position to produce an efficient and testable design.

To be able to represent the constraints posed on the design by the individual cells efficiently, we have adopted the same graph approach as used by Abadir (Abadir and Breuer, 1985 and Breuer *et al*, 1988). In this graph, the nodes represent the cells and the arcs the interconnection between cells. The graph model is hierarchical in that a node can represent a fully characterized design (as opposed to a cell) which itself can be represented by another subgraph. Figure 3 shows the graph representation of the example circuit in Figure 2. Various attributes are attached to each node to represent the cell characteristics. Typical node attributes for the graph in Figure 3 are presented in Table 1.

Node name:	PLA	Node name:	R1
Type:	combinational	Type:	register
Order:	simple	Order:	simple
Design style:	PLA	Input ports	(i1,8); (clk,1
Input ports:	(i1,8)	Output ports:	(out,8)
Output ports:	(out,8)	i-mode:	yes
i-mode:	yes (1-1)	Mode:	load
Test vectors:	yes	Size:	8

Table 1. Node Attributes

Most of the node attributes are easily understood. The *order* label is used to judge the i-mode propagation characteristics of the cell. If the node requires a complex procedure to activate the i-mode (e.g. a shift register), it is given an *intermediate* order factor. In the case of simple cells like adders, word-level logic gates, multiplexers etc., the order factor is *simple*. If a cell does not have an i-mode the order factor is *complex*. If the *i-mode* field is *yes*, various attributes are attached to represent the activation conditions as illustrated in Figure 1. The *test vector* field is used to specify the availability of test vectors for the node or the test strategy.

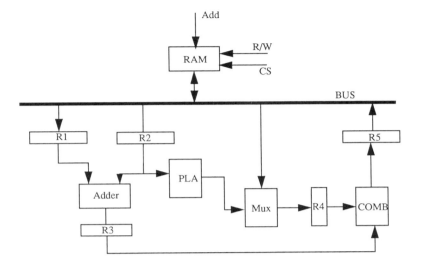

Figure 2 Example circuit

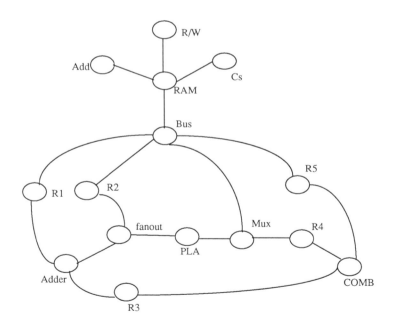

Figure 3 Graph representation of example circuit

It must not be assumed that the arcs in the graph model of the design represent only the interconnection between cells. In fact, arcs also have various attributes to represent the testability values of the busses in the design. Typical attributes are presented in Table 2.

Arc-name:	(RAM:Bus)
Controllability:	c
Observability:	o
Type:	bidirectional

Table 2. Arc Attributes

The *arc-name* indicates the nodes that are connected and the *type* field indicates the directions of the arc: *unidirectional* or *bidirectional*. The controllability and observability estimates are used for the path planning exercise as will be shown later as well as in the design partitioning as explained above.

7.2.4.1 Test program generation.

Test program generation involves selecting a target, creating suitable paths for fault sensitization (*S-path*) and test vector propagation (*T-path*). This creates a subgraph which can be analyzed symbolically. As an example, suppose we need to create a test program for the node *COMB* in Figure 3. To compute the S-path to the primary output, the system selects the first node (nodes) connected to the output of the target. In the present situation the only node is *R5*. It next checks the type of node. If it is a primary output the procedure is stopped. If not, it finds the node (nodes) connected to the output of the present node. It uses the bus observability estimates to judge the most promising path to trace. If the present node is of type fanout, the path with the highest observability estimate is traced. This procedure is repeated until the primary output of the design is reached. So for the present situation the S-path is *COMB, R5, Bus*. A similar approach is used to create the T-path for every input bus of the target. The testing subgraph for the node *COMB* is presented in Figure 4.

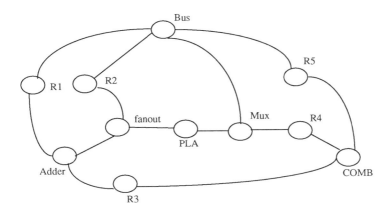

Figure 4 Testing subgraph for node COMB

The next stage is to analyze this subgraph to show the life cycle of each test vector applied to the target. We term this analysis *creating a test program for the target*. To reduce the amount of computation involved, this analysis is handled symbolically. This results in a significant improvement over the vectorized representation of test vectors (Murray and Hayes, 1988, Njinda and Moore, 1991 and Alfs *et al*, 1988). To create a test program for node *COMB*, the first step is to assign symbols to represent the set of test vectors at the input and output arcs of the target. A time-frame is used to represent the initializing conditions. If the node is of type combinational, we assume a zero delay and set the required symbols at the input/output arcs at time-frame zero. In the case of register nodes, the output arcs are given a time-frame m; where m is the number of clock periods required before the output arc is completely specified. *COMB* is of type combinational and so the inputs are given the values $s1$ and $s2$ while the output takes the value z all at time-frame 0. Using the i-mode activation conditions of all the nodes in the subgraph of Figure 4, the test program for node *COMB* is presented in Table 3, where all times have been normalized to one.

1	Bus(s1), Mux(select Bus-Mux), R4(load)
2	Bus(s2), R1(load)
3	Bus(0), R2(load), Adder(-)
4	R3(load), COMB(-)
5	R5(load), Bus(z)

Table 3. Test Program for Node *COMB*

If the node *COMB* contains T test vectors, the cycle in Table 3 has to be repeated T times. This is as illustrated in Table 4 for two successive vector, $t_{1,1}, t_{1,2}, r_1$ and $t_{2,1}, t_{2,2}, r_2$.

1	Bus($t_{1,1}$), Mux(select Bus-Mux), R4(load)
2	Bus($t_{1,2}$), R1(load)
3	Bus(0), R2(load), Adder(-)
4	R3(load), COMB(($t_{1,1}, t_{1,2}, r_1$)
5	R5(load), Bus(r_1)
6	.... Bus($t_{2,1}$), Mux(select Bus-Mux), R4(load)
7	.... Bus($t_{2,2}$), R1(load)
8	.... Bus(0), R2(load), Adder(-)
9	.... R3(load), COMB($t_{2,1}, t_{2,2}, r_2$)
10	.... R5(load), Bus (r_2)

Table 4. Test Process for Node *COMB*

So the total test time for node *COMB* is $5T$. If the conflict between the resources can be fully analyzed and *no-ops* are used, the test time for a node can be reduced (Abadir and Breuer, 1985).

7.2.4.2 Design for testability modifications.

The process discussed in the previous section creates a test program for nodes for which it is possible to establish the required paths for vector propagation and justifications. Here we discuss how the system can use multiplexers and BILBO structures to modify the design if the necessary paths cannot be established. Rules for embedding these structures in the design have been discussed by Breuer (Breuer *et al*, 1988) where only BIST is considered. These rules give the cost of using any structure in terms of the area overhead, number of extra input/output pins, testing time and ease of implementation. However, their approach is not applicable here since we are interested in adding extra hardware only when analytic testing is not successful. That is, the approach used here is driven by the failures obtained when generating test vectors analytically. To determine when it is necessary to utilize these test structures and the sort of structure needed, we have implemented four different rules in the if-then format. (Note: An offending node is a node without an i-mode).

1) *If* the designer prefers to use a specific test method for the target, *then* use the stated method and determine the cost measures.

2) *If* there are one or more offending nodes and test vectors for the target are not available and the target is of type combinational, *then* use a BILBO structure.

3) *If* it is not possible to establish a T-path and/or an S-path and test vectors are available for the target and the total number of offending nodes is 1, *then* use a multiplexer structure to bypass the node.

4) *If* it is not possible to establish both a T-path and an S-path and the number of offending nodes is greater than 1 and test vectors for the target are available *then* use BILBO or scan path techniques.

Rule 1 is meant for the experienced designer who by just looking at a design is able to decide which test strategy is best or if at certain nodes it will be advisable to use a specific method. In Rule 2, it is possible that there are no test vectors for the target or no available test strategy. Also the number of offending nodes might be greater or equal to one. If the node is of type combinational, then the best solution would be to embed a BILBO structure for testing that node. In Rule 3, if the number of offending nodes is equal to 1, then the system will always suggest that a multiplexer structure be used. Where there are many offending nodes (Rule 4), it is assumed that it is impractical to place a multiplexer structure at every interface. It is best in such situations to use a scan path (if test vectors are available) or a BILBO technique. However, if the costs associated with using these techniques are not acceptable to the designer, by using the failure information, the system would be able to direct the search towards a more suitable solution or exit without a solution.

7.2.4.3 An example

To illustrate the procedure for embedding test structures into designs, consider creating test programs for all the nodes in Figure 3. Nodes *Add*, *PLA*, and *Mux* cannot be tested due to the fact that no i-mode is available in node *COMB*. That is, node *COMB* is an offending node in each case. Hence it is not possible to propagate the test response vectors to observable outputs. To solve the problem in each case a multiplexer structure is required for bypassing the node *COMB* in the test mode. However, we solve the problem for the general case using the following steps. (Note: in the present implementation only multiplexer structures are considered).

1) Find all nodes for which a test program cannot be created.

2) Find the set of offending nodes. That is if thereare m nodes for which test programs cannot be created and the set of offending nodes in each case is O_i, then the set of offending nodes for the design is $O_1 \cup O_2 \cup ... \cup O_m$.

3) Add the minimum number of multiplexer structures such that each node is testable (that is, there is an S-path from each output bus of the target to the primary outputs and a T-path to each input port of the target and none of the paths have conflicts with each other).

In the present example, the only offending node is *COMB*. For the design to be testable two 2-input multiplexers have to be added to the design as illustrated in Figure 5. The cost for making the design testable is the area required for the addition of two multiplexers and two extra input/output pins. One is required to select the test/normal mode and the other selects which of the nodes is being tested.

As designs become increasingly complex, it is common to have cells that have two modes of operation. In the normal mode they perform their normal function. In the test mode certain cells are bypassed such that test data can be transferred unchanged from the primary inputs to the target and test responses are propagated unchanged to the primary outputs. A cell that can function in both modes is called a testable structure. The

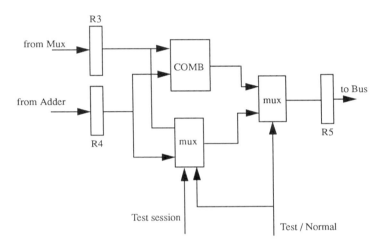

Figure 5 Embedding the multiplexer structure in a design

hardware overhead when using this approach is very high, since every cell is assumed to be opaque and extra hardware has to be added to make it transparent in the test mode. This is certainly not the case if each cell can be analyzed to determine its characteristics as we do in HIT. Modifications only need to be made when the whole design has been considered in such a way that the least amount of hardware is added to make every node testable.

7.3 Conclusions

The general trend in industry is to get designs right first time. This philosophy needs to be extended to testing. HIT, currently being developed at the University of Oxford, falls under a new breed of test generating systems that will ensure that this is achieved. It not only provides efficient test generation but can also guide the designer through various design for test modifications. Any extra circuitry is targetted at the correct location thereby reducing cost, hardware overhead and the effect on circuit performance. It must not be assumed that the modifications suggested by HIT are the only ones or that they are necessarily the best. The idea of designing such a system is to try and create a dialogue between the designer and the test expert. By giving the designer expert knowledge about the areas of his design where testing would be difficult he is always in a better position to suggest alternative solutions. Also conferring with the designer is more important than using arbitrary measures to improve the testability of a circuit. Finally, even though the designer is in the best position to modify the design to improve testability, having a system such as HIT to guide him/her will reduce the effort required and in most cases improve the final product.

7.4 References

ABADIR M.S. AND BREUER M.A., "A Knowledge-based System for Designing Testable Chips", IEEE Design & Test, Aug. 1985, pp. 56-68.

ABADIR M.S., "TIGER: Testability Insertion Guidance Expert System", ICCAD 1989, pp. 562-565.

ALFS G., HARTENSTEIN R.W. AND WODTKO A., "The KARL/KARATE System: Automatic Test Pattern Generation on RT-Level Descriptions", ITC 1988, pp. 230-235.

BENNETTS R.G., "Designing Testable Logic Circuits", Addison-Wesley, 1984.

BHATTACHARYA D. AND HAYES J.P., "A Hierarchical Test Generation Methodology for Digital Circuits", Journal of Electronic Testing: "Theory and Applications", No. 1, 1990, pp. 103-123.

BREUER M.A., GUPTA RAJIV AND GUPTA RAJESH, "A System for Designing Testable VLSI Chips", IFIP workshop on Knowledge Based Systems for Test and Diagnosis, Grenoble, France, Sept. 1988.

FUJIWARA H. AND SHIMONO T, "On the Acceleration of Test Generation Algorithms", IEEE Trans. Comput., Vol. C-32, No. 12, Dec. 1983, pp. 1137-1144.

GOLDSTEIN L.H., "Controllability/Observability Analysis of Digital Circuits", in Tutorial, Test Generation for VLSI Chips, V. D Agrawal and S. C. Seth, IEEE 1987, pp. 262-270.

ILLMAN R.J., "Design of Self Testing RAM", Silicon Design Conf. 1986, pp. 439-446.

JONE W.B. AND PAPACHRISTOU C.A, "A Coordinated Approach to Partitioning and Test Pattern Generation for Pseudoexhaustive Testing", Design Automation Conf., June 1989, pp. 525-530.

KIM K., TRONT J.G. AND HA D.S., "Automatic Insertion of BIST Hardware Using VHDL", 25th ACM/IEEE Design Automation Conference, 1988, pp. 9-15.

LENSTRA J. AND SPAANENBURG L., "Using Hierarchy In Macro Cell Test Assembly", ETC 1989, pp. 63-70.

MARNANE W.P. AND MOORE W.R., "A Computational Approach to Testing Regular Arrays", Int. Conf. on Systolic Arrays, Killarney, Ireland, May 1989.

MELGURA M., "Fault Simulation at the Functional Level", NASI Testing and Fault Diagnosis of VLSI and ULSI, Digest of papers, Italy 1987.

MICALLEF S.P. AND MOORE W.R., " Hierarchical Path Sensitization for Testability Analysis and DFT", IEE Colloquium on *Design for Testability*, London, May 1991.

MURRAY B.T. AND HAYES J.P., "Hierarchical Test Pattern Generation Using Pre-computed Test Modules", ITC 1988, pp. 221-229.

NJINDA C.A. AND MOORE W.R., "Acceleration of ATPG by Making use of Hierarchy", CAD Accelerators, A. P. Ambler, P. Agrawal and W. R. Moore (Eds.), Elsevier 1991, pp. 115-130.

ROTH J.P., "Diagnosis of Automata Failures: a Calculus and a Method", IBM Journal of Res. and Dev., Vol. 10, 1966, pp. 278-281.

SALICK J., UNDERWOOD B., KUBAN J. AND MERCER M.R., "An Automatic Test Pattern Generation Algorithm for PLAs", ICCAD 1986, pp. 152-155.

SHEN L. AND SU S.Y.H., "A Functional Testing Method for Microprocessors", FTCS 1984, pp. 212-218.

SOMENZI F. AND GAI S., "Fault Detection in Programmable Logic Arrays", Proc. IEE, Vol. 74, No. 5, May 1986, pp. 655-668.

SRINIVISAN R., NJINDA C.A. AND BREUER M.A., "A Partitioning Method for Achieving Maximal Test Concurrency in Pseudo-exhaustive Testing", VLSI Int. Test Symposium, April 1991, New Jersey.

THATTE S.M. AND ABRAHAM J.A., "A Methodology for Functional Level Testing of Microprocessors", FTCS 1978, pp. 90-95.

THEARLING K. AND ABRAHAM J., "An Easily Computed Functionality Measure", ITC 1989. pp. 381-390.

URDELL J.G AND MCCLUSKEY E.J., "Efficient Circuit Segmentation for Pseudo-Exhaustive Test", ICCAD 1987, pp. 148-151.

VILLAR E. AND BRACHO S., "Test generation for Digital Circuits by Means Of Register Transfer Languages", IEE Proc., Vol. 134, Pt. E, No. 2, March 1987, pp. 69-77.

Chapter 8
Use of Fault Augmented Functions for Automatic Test Pattern Generation

N. Jeffrey, G.E. Taylor, B.R. Bannister and P. Miller

8.1 Introduction

Techniques for the automatic generation of test patterns for digital circuits may be classified into four main areas;

1) random pattern generation ... random input sequences which may be exhaustive, pseudo-exhaustive or used in conjunction with fault simulation tools to assess fault coverage.

2) heuristics ... these include checkerboard patterns and walking ones and zeros as well as functional tests; for automatic generation of such sequences there is the implication of some kind of expert system and the need for simulation to assess fault coverage.

3) path sensitisation ... this encompasses the many variations on the D-algorithm and is probably the most widely used algorithmic technique.

4) algebraic techniques.

It is with the last that this paper is concerned. The most widely known algebraic technique is that of Boolean Difference in which boolean functions $G(\underline{X})$ and $F(\underline{X})$, describing the fault free and faulty circuits respectively, are produced. A test function $T(\underline{X})$ is formed from the exclusive-OR of $G(\underline{X})$ and $F(\underline{X})$ and clearly, from the definition of the exclusive-OR operator, any set of input values $\underline{X}_t$ which makes $T(\underline{X})$ take the value 1 is a test for the fault considered. The technique can be used for any fault model which affects the logic of the circuit (ie it is not restricted to 'stuck-at' faults) and it produces all patterns which detect the particular fault embedded in F($\underline{X}$). However it is at a considerable disadvantage compared with other techniques in that it deals with only a single fault at each pass and requires extensive symbolic manipulation. An alternative approach was first presented in (Miller and Taylor, 1982) and will be explored and extended in this paper.

8.2 *The Fault Augmented Function (FAF)*

A fault augmented function (FAF) describes the function realised by a good logic circuit and the functions realised by many faulty circuits simultaneously. Each fault that is to be injected into the function has a boolean variable associated with it that is 1 when the fault is present, and 0 when it is not. This technique is based on the work of Poage (Poage, 1963). The number of fault variables injected into the function can be kept to a minimum by selecting a single fault to represent each fault equivalence class of interest. As an example, consider the AND gate shown below.

$$A \longrightarrow \boxed{\&} \longrightarrow Z=A.B$$
$$B \longrightarrow$$

Injecting the faults F_1 = A stuck-at-1, F_2 = B stuck-at-1 and F_3 = Z stuck-at-0 gives :-

$$Z_{(augmented)} = (A + F_1).(B + F_2).\overline{F_3}$$

Alternatively, injecting the fault F_4 = short between A and B, modelled as wired-OR, gives :-

$$Z_{(augmented)} = (A + F_4.B).(B + F_4.A)$$

Injecting all four faults gives :-

$$Z_{(augmented)} = ((A+F_1) + F_4.(B+F_2)).((B+F_2) + F_4.(A+F_1)).F_3$$

8.3 *The Original Algorithm*

The original algorithm proposed for use with the FAF formed $T(\underline{X},\underline{F})$, the exclusive-OR of Z (the original circuit function) and Z(augmented) (the fault augmented function), and expressed this in sum of products form. Each product term then represents a test for one of the modelled faults and may be interpreted as a product of input variables, X_i, (defining the test pattern), true pseudovariables, F_i, (defining the fault under test) and false pseudovariables, $\overline{F_i}$, (defining the faults which invalidate this particular test). For example the term $X_1\overline{X_2}X_3F_1F_3\overline{F_2}\overline{F_{14}}$ indicates that the input pattern $X_1=1$, $X_2=0$, $X_3=1$ detects the multiple fault comprising the single faults F_1 and F_3 occurring simultaneously provided that neither fault F_2 nor fault F_{14} is also present. It should be noted that the algorithm produces only those multiple faults for which there exists a test to distinguish them from their constituent single faults. This test function is a superset of the fault dictionaries produced by other methods and, using suitable heuristics, tests may be selected from it to produce appropriate sequences for go/no-go testing or for fault diagnosis.

8.4 *Generating the FAFs*

Before the test set can be derived, a method of deriving the FAFs for a circuit described in terms of a directed graph is needed. There are several ways of tracing through a circuit, of storing the boolean functions derived, and of manipulating those functions into the form required. This section explains the methods used in this project and the reasoning behind them.

The most convienient form for the test function which is ultimately required is a sum of products form, because it can be directly translated into a set of test patterns. A sum of products boolean function can be stored as a linked list of terms, where each term is an array of bits. Each variable on which the function depends is represented by two bits in the array, one which is set when the true form of the variable is present in the term, and one which is set when the inverted form is present. This representation allows efficient manipulation of the functions using bitwise operations which can be performed very quickly by a computer.

Because of this it was decided that this representation would be used throughout the FAF generation process. The FAF for each circuit net is stored in this way, starting with the primary inputs (whose FAFs can be derived immediately). The FAF for each gate output can then be derived from the FAFs of its inputs and the gate's logic function. This process is recursive, evaluating the function of a net and then calling itself on the output of the gate driven by that net, provided that all the gate's other inputs have also been evaluated. Using this method, a simple loop to generate the FAFs for each primary input will end with the FAFs for every net in the circuit being derived, and producing the required functions for the primary outputs. An integral part of this process injects faults into the functions as it works through the circuit. The faults to be injected are specified in a fault list which is read in by the program at start up. Each fault type (stuck-at-0, stuck-at-1, wired-AND, wired-OR, etc) that is recognised by the program has a specific piece of code associated with it which injects the relevant fault variables into the functions of the affected nets.

Clearly it is unnecessary to store the FAF for every net in the circuit. All that is ultimately required are the FAFs for the primary outputs. The above process allows for this by releasing the storage used for any FAF as soon as it is no longer required (ie as soon as the output of the gate which it drives has been evaluated). Using the method described above for storing the FAFs this is very easy to do. A large array of terms is allocated initially, along with a LIFO stack of pointers to unused terms which is initially full (containing a pointer to every term in the array). Whenever a new term is required the stack is popped and the corresponding term is used. When a FAF is no longer needed it is released by simply running through the linked list of terms which represent it and pushing pointers to each onto the stack. This means that these released terms will be re-used immediately, filling in the holes in the memory space and removing the need for garbage collection to prevent excessive memory consumption.

Working forwards through the circuit in this way was preferred because it only visits each node in the circuit once. The function of a node is evaluated and then used to evaluate all the nets which it drives, before being released. In a circuit with several outputs which contains fanout, the trees which drive each primary output will overlap. Hence a simple tree search which works backwards from the primary outputs will evaluate the overlapping sub-trees more than once, which is inefficient. With this intention, the use of sum of products form throughout the FAF evaluation process is sensible because it allows for efficient and simple memory management, and precludes the need to convert another FAF representation into sum of products form at the end of the process.

During the evaluation process FAFs must be ANDed together, ORed together and inverted. Each of these operations will produce a new FAF which may contain redundant terms. For example, a term which contains both the true form of a variable and its inverted form is identical to zero and can be removed from the function. It is desirable to perform as much reduction as possible, as soon as possible, during FAF generation. This is because longer FAFs will be more time-consuming to process at the next stage and will result in even longer and more redundant ones. It was originally intended that the set of reduction rules proposed by Zissos (Zissos, 1976) would be used to reduce each FAF as soon as it was derived, or in the process of deriving it. However, applying these reduction rules is time-consuming in itself, being the most computationally expensive part of the algorithm originally implemented by Miller (Miller and Taylor, 1982). With each term in a FAF stored bitwise some of these reduction rules can be implemented very efficiently (eg checking for the presence of both a variable and its inverse). Consequently it was decided to implement a sub-set of the Zissos rules; specifically :-

$$A.0 = 0 \quad A + 1 = 1 \quad A.\overline{A} = 0 \quad A + A.B = A$$

These rules can be implemented relatively easily and remove redundant terms that are likely to occur during normal FAF generation. The redundant terms that they do not remove should not occur in a circuit that does not contain redundant logic. It was assumed that such redundant logic would be largely removed as part of the design process, since it is undesirable for many other reasons, and that which remained would be minimal and hopefully, therefore, would cause few problems. This was found to be the case with a few small sample circuits (ie without the extra reduction rules, redundant terms did not appear in the FAFs for the primary outputs) but this approach has not been fully evaluated and is the subject of review.

Another aspect of FAF generation using this method is that of whether to generate the FAF or its inverse for any particular net. Because the FAFs are stored in sum of products form the most computationally complex operation to perform on them is inversion. Hence, avoiding unnecessary inversions is important. This is achieved by producing either the FAF or its inverse for any particular net, depending on which is easier. For example, if the output of a NAND gate is being evaluated it can be done using either,

$$Z = \overline{A.B} \quad \text{or} \quad Z = \overline{A} + \overline{B}$$

If both inputs are represented by FAFs in the true form the simplest thing to do is to evaluate A.B and store this as the inverse FAF of the result. Conversely, if both inputs are represented by their inverse FAFs it is easier to generate the true form of the result (since this involves only a single OR operation). If one input is represented by its true FAF, and the other by its inverse FAF the situation is more complicated. At present the algorithm would invert the true FAF, and use the OR operation to generate the true FAF of the output. This is because ORing together FAFs is much simpler than ANDing them (it is basically a question of concatenating two linked lists).

8.5 Problems with FAF Generation

The method described above for generating the FAFs of the primary outputs of a combinational logic circuit, subject to single stuck faults and multiple line short circuits modelled as wired-OR or wired-AND, was implemented in the C programming language. It was found that even for small circuits (less than 10 gates) FAF generation was very slow (considerably longer than it took to generate a complete single stuck fault detection test set using the D-algorithm for the same circuit), and for larger circuits FAF generation not only took considerable CPU time, but also consumed a large amount of memory.

One particular example consisted of a circuit containing 8 gates, with 4 inputs and 1 output, and a collapsed fault list of 25 single stuck faults.

FAF generation required 42 CPU seconds on a VAX 11/750 and resulted in 904 terms, which appears to be rather excessive. But, considering that the FAF for the primary output is a function of 4 input variables and 25 fault variables it could, in the worst case for a sum of products form, contain 2 to the power 28 terms (approx 268 million). In this context 904 does not seem too bad. For the larger circuits mentioned above, FAF generation was practically impossible because of the large number of terms in the result. However, generating the good circuit function (using exactly the same program, but supplying it with an empty fault list) proved to be very fast in many cases (at least an order of magnitude faster than DALG test generation). The problem, therefore, is how to cope with the large number of fault variables, which will in general greatly exceed the number of input variables. Considering the above example again, since the algorithm at this point makes no distinction between a fault variable and an input variable, introducing 25 fault variables is actually equivalent to introducing 2 to the power 25 possible fault situations (approx 33 million) rather than the intended 25. In order to cope with the manipulation of functions of such a large number of variables some constraint must be imposed on the system.

It was decided that a sensible approach would be to make the single fault assumption, ie at any one time only one of the faults being considered can be present. For FAF generation, however, this does not mean that only single faults can be considered. It simply means that only one fault variable will be true at any one time. There is no reason why a fault variable cannot explicitly represent a multiple fault. What this assumption effectively does is to stop the algebra from automatically taking into account all possible multiple faults (of all multiplicities), and instead allows particular multiple faults to be specified for consideration. This allows much greater control over the algorithm and leads to greatly improved performance of the FAF generation process, both in terms of CPU time and memory requirement.

8.6 *Implications of the Single Fault Assumption*

This section explains the implications of the single fault assumption for FAFs and their manipulation. Firstly, consider the form that a FAF

170 Use of fault augmented functions

might have (regardless of what it represents). It might look something like the following;

$T_1 + ...$
$T_{(x-1)} +$ Terms that are always present.

$T_x.\overline{F_i}.\overline{F_j}... + ...$
$T_n.\overline{F_q}.\overline{F_p}... +$ Terms present in the good function, but removed by certain faults.

$FT_1.(F_a+F_b+...) + ...$
$FT_m.(F_c+F_d+...)$ Terms present in various faulty functions that are not present in the good function

This representation seems intuitively sensible because any function (represented by the terms T_1 to T_n) can be changed into any other by the presence of a fault. Any terms in the good and faulty functions would remain unchanged, any terms in the good function that do not appear in the faulty function have a relevant $\overline{F}$ attached to them, and any terms that appear in the faulty function but not in the good function are added to the list with an appropriate true F attached. As many faults as required can be introduced into the function in this way to yield a FAF that represents the good function and all modelled faulty behaviours.

The above form is not quite sum of products (because of the FT(F+F+...) terms), but it is very similar. If the FAF were represented by a true sum of products form (as it was originally) it could contain two types of term not included in the above representation. They are;

1) terms which contain more than one true F variable. These terms, however, are equal to zero and can be removed because the single fault assumption means that at least one of the F variables would always be zero, hence the whole term would always be zero.

2) terms which contain both true Fs and inverted Fs. If there is more than one true F case 1 applies. If there is only one true F and one or more inverted Fs then the inverted Fs can simply be removed from the term. To justify this consider the two cases where the true F variable is 0 and 1. If it is 0 the whole term is 0. If it is 1 it must be the only F that is 1 (given single faults) hence all the

inverted Fs would be 1, and the term would definitely be present. This is exactly the same behaviour as the same term with all the inverted Fs removed.

Given this representation it is clear that terms can be stored much more efficiently. Firstly, because no term contains both true Fs and inverted Fs, the size of each term can be reduced because it is no longer necessary to have two bits per F variable. Instead there needs to be one bit per F variable and a single additional control bit to indicate how to interpret the F bits (ie it is a true type term or an inverted type term). Because F variables generally outnumber primary input variables by a considerable factor (eg a sample circuit containing random logic had 15 inputs and a collapsed fault list containing 159 faults) this will very nearly halve the size of each term, and hence the total memory required.

Secondly, there will be considerably fewer terms to store. All terms with more than one true F will disappear, and, given the above storage method, many terms containing a single true F can be merged into one. The control bit would then indicate not only that the F bits represented true F variables, but also that they are ORed together rather than ANDed. The reduction in the worst case number of terms because of this is considerable. Given n inputs and m faults, the worst case number of terms for the original sum of products representation is $2^{(n+m-1)}$. In the new form, however, the worst case number of terms is 2^n+1. That is, the worst case is now independent of the number of faults and is only twice that of the worst case number of terms for the good function on its own.

It was mentioned earlier that the most convienient form for the test functions (which are derived from the primary output FAFs and have a similar form) is a sum of products form. This form has now been sacrificed slightly, but this is not a disadvantage. In fact the new form is even better. The test functions would now be expected to take a similar form to the FAFs, except that they would only contain terms of the form FT(F+F+...). This would then be interpreted as a test pattern, as defined by FT and a list of faults that would be covered by that test. This is a much more compact form than the original (which would have needed several terms to represent each term of the above form) and provides the information required in a more readily accessible form.

The single fault assumption clearly leads to a more compact representation for FAFs, which reduces the memory requirement of the FAF generation process. The reduction in the number of terms should also lead to faster processing on the assumption that shorter FAFs will take less time to manipulate (consider the problem of inverting a FAF - the fewer terms the better). However, the new form for FAF representation can be taken advantage of directly to speed up processing even further.

Consider, for example, the problem of inverting a single term. Previously this would have resulted in a list of terms each containing a single variable which was the inverse of that which appeared in the term being inverted. With the new form, the same operation must be carried out for the portion of the term representing the primary input variables (which has not changed at all) but once this has been done all that is left is a single term containing no input variables, with exactly the same fault variables as the original term, only interpreted the opposite way (ie simply toggle the control bit). That is :-

$$\overline{A.B.\overline{F_1}.\overline{F_2}.\overline{F_3}} = \overline{A} + \overline{B} + (F_1+F_2+F_3)$$

$$\overline{A.B.(F_1+F_2+F_3)} = \overline{A} + \overline{B} + \overline{F_1}.\overline{F_2}.\overline{F_3}$$

Clearly the overhead of injecting fault variables is much less than before, and is much less dependent on the number of faults injected (in the above example it makes no difference how many faults are present).

Now consider ANDing two terms together :-

$$A.B.(F_1+F_2+F_3).C.D.(F_3+F_4+F_5)$$
$$= A.B.C.D.(F_1+F_2+F_3).(F_3+F_4+F_5)$$
$$= A.B.C.D.(F_3)$$

(since $F_x.F_y=0$ unless x=y)

Compare this to the original manipulation :-

$$(A.B.F_1 + A.B.F_2 + A.B.F_3).(C.D.F_3 + C.D.F_4 + C.D.F_5)$$

$$= A.B.C.D.F_1.F_3 + A.B.C.D.F_1.F_4 + A.B.C.D.F_1.F_5 +$$
$$A.B.C.D.F_2.F_3 + A.B.C.D.F_2.F_4 + A.B.C.D.F_2.F_5 +$$
$$A.B.C.D.F_3 \quad + A.B.C.D.F_3.F_4 + A.B.C.D.F_3.F_5$$

$$= A.B.C.D.F_1.F_4 + A.B.C.D.F_1.F_5 +$$
$$A.B.C.D.F_2.F_4 + A.B.C.D.F_2.F_5 +$$
$$A.B.C.D.F_3$$

The first case above, using the new form, can be performed extremely easily. It involves a bitwise OR between the bits representing the input variables, and a bitwise AND between the bits representing the fault variables. The second case (the original form) is considerably more complex. It involves a total of 9, two-term AND operations performed by bitwise OR between the two terms. Remember also that the original form had almost twice as many bits, hence twice as many words to OR together. This is followed by the application of reduction rules to remove unnecessary terms (which is very time-consuming). The second form also results in five terms instead of one, four of which represent information on multiple faults and could be dropped, but which would have remained in the original form causing more computational complexity in later stages of the FAF generation process.

The reduction rules used to remove unnecessary terms can also be modified to take advantage of the new representation. It is possible to perform some form of reduction whenever there is a sub-set or equality relationship between the primary input variables of two terms that are ORed together (Exactly what happens depends on the types of the two terms and which one is PI sub-set). Previously, sub-set relationships had to be evaluated for the complete terms (fault variables as well).

The above examples illustrate the advantages of the new form. The general trend that is followed by all the new manipulation rules is that the number of operations to be performed is not significantly dependent on the number of fault variables present. This satisfies the original intention, which was to make FAF generation possible in a similar time to good function generation.

Having implemented the new FAF representation form and new manipulation rules a great improvement in performance was achieved. In one example, a circuit which could not be handled before, as FAF generation ran out of memory after several days of CPU time, but whose good function could be generated in 0.36 CPU seconds (on a VAX 11/750), was used. The circuit had 15 inputs, 20 outputs, 74 gates, 89 nets and a collapsed fault list containing 159 single stuck faults. Using the new functions FAF generation was completed in 5.71 CPU seconds. This compares to 9.54 seconds for DALG test generation using the same fault list.

8.7 Extracting Tests from FAFs

Once the FAFs have been generated the good functions can be extracted from them by simply copying the primary input bits of each inverted type term and zeroing all the fault variable bits (although even this is not strictly necessary since the data stucture representing the FAF can also represent the good function, it is merely interpreted in a different way). The test functions can then be generated by XORing the FAFs with the good functions. The resulting test functions then consist wholly of terms of the form $A.B.C...(F_1+F_2+F_3+...)$, each of which represents a test pattern and the faults covered by it. The test function cannot contain inverted type terms because that would imply that the test function could be forced to 1 in the absence of all faults (ie that some test pattern could distinguish between the good circuit and itself, which is impossible). The test functions contain all tests and their fault coverage. Any required test set can be extracted from them very easily.

A full symbolic XOR is a time-consuming operation and it was hoped that a method could be found to extract any required tests directly from the FAFs, without having to generate the test functions at all. That is, the post processes that would be performed to extract a particular test set from the test functions could be modified to operate directly on the FAFs and extract the required tests from them. The test function is formed as follows:-

$$X = G \text{ XOR } F = G.\overline{F} + \overline{G}.F$$

where :-

X is the test function
G is the good circuit function
F is the fault augmented function

G and F are of the form :-

$$G = T_1 + T_2 + \ldots + T_i + \ldots + T_n$$

where each T is a product term of primary input variables (eg $A.\overline{B}.D$)

and :-

$$F = T_1.\overline{F_a}.\overline{F_b}\ldots + \ldots + T_n.\overline{F_x}.\overline{F_y}\ldots + FT_1.(F_i+F_j+\ldots) + \ldots + FT_m.(F_p+F_q+\ldots)$$

where each FT is a product term of primary input variables and each F is a single fault variable.

Hence :-

$$G.\overline{F} = (T_1 + \ldots + T_n).(\overline{T_1} + F_a + \ldots)\ldots(\overline{T_n} + F_x + \ldots).$$

$$(\overline{FT_1} + \overline{F_i}.\overline{F_j}\ldots)\ldots(\overline{FT_m} + \overline{F_p}.\overline{F_q}\ldots)$$

and :-

$$\overline{G}.F = (\overline{T_1}\ldots\overline{T_n}).(T_1.\overline{F_a}\ldots + \ldots + T_n.\overline{F_x}\ldots +$$

$$FT_1.(F_i+F_j+\ldots) + \ldots + FT_m.(F_p+F_q+\ldots))$$

$$= (\overline{T_1}\ldots\overline{T_n}).(FT_1.(F_i+F_j+\ldots) + \ldots + FT_m.(F_p+F_q+\ldots))$$

Extraction of a test for a specific fault (or group of faults) from the first part of the test function can be done as follows. From each inverted type term in the FAF (which must now be interpreted as a product of sums term with the sense of each variable present being reversed) which does not contain the fault variable required (or all fault variables in the case of a group), select a primary input variable to be ANDed with the partially formed test term that does not force it to zero (because its inverted form already appears). If possible choose a variable already present in the test term. Do the same for each true type term (the FT terms above) that contains any of the fault variables for which the test is being derived. Finally, select one term from the good function that can be ANDed with the test term without yielding zero. The decisions made during this process can be guided if assignments to particular primary input variables are especially desirable (or undesirable). The operations involved in this process are computationally simple because of the bitwise form in which the terms are stored.

The second part of the test function simplifies immediately because of the relationship between the good function and the FAF. Extracting a test from this simplified expression can be done by finding a true type term in the FAF which contains the relevant fault variable (or variables) and then selecting one primary input variable from each of the inverted type terms which will not force the test term to zero, inverting it, and then ANDing this (via a bitwise OR) with the test term being derived.

If at any point, in either of the above two processes, it is impossible to select a variable that does not force the test term to zero (because the test term already contains its inverse) then earlier choices must be changed if a test is to be derived. Both problems are effectively that of satisfying a product of sums formula (which itself is an NP-complete problem (Cook, 1971)). Note, however, that the formula to be satisfied is effectively only a function of the primary input variables, not of the fault variables as well.

8.8 Current State of the Software

The current implementation of the FAF test generator is written in C and has been ported to a VAX 11/750 running VMS V4.5, and to a SUN SPARCstation 1 running SUNOS 4.1.3c. The program is currently capable of handling combinational circuits described in terms of AND, NAND, OR and NOR gates with arbitrary numbers of inputs, NOT gates, 2-input XOR and XNOR gates, and single-input, multi-output buffers. It is capable of injecting single stuck faults and short circuits involving two or more nets, which can be modelled as wired-AND or wired-OR.

A net which fans out is replaced by a net (representing the trunk of the fanout net) driving a multi-output buffer which drives several other nets (each one representing a branch of the fanout net). This means that the trunk and each branch of a fan out net can be treated as a separate fault site if desired. That is, faults injected onto the trunk affect the FAF of the trunk, all branches, and all FAFs derived from them. A fault injected onto a specific branch only affects the FAF of that branch and the FAFs derived from it, it does not affect the trunk or any other branch of the fan out net.

The software currently reads a circuit description and a fault list and then evaluates the FAFs and inverse FAFs for each primary output in the presence of all faults in the list. It then generates the good function for each primary output (extracted directly from the FAF), the inverse of the good function (extracted directly from the inverse FAF), and finally the test function.

At present the program derives the test functions from the FAFs by performing a full symbolic XOR. This is for program testing only, it is not intended that this should be used as the ultimate method of deriving the tests. It does not extract any particular test set from the test functions. Rather, it writes out the complete test function for each primary output in a human readable form that looks like a boolean function.

8.9 Appraisal of the FAF Technique

The FAF technique does not compete directly with other combinational test generation algorithms such as DALG, PODEM, FAN etc, because it requires more CPU time and memory. However, it generates more information than these techniques, providing details of all useful tests and the faults that they cover. This information can be used to derive different types of test set (eg go/no-go or fault location) and/or alternative test sets for situations where it is not possible to apply particular tests (eg for sub-modules within a larger design).

FAF test generation is not restricted to single stuck faults. Other techniques can also theoretically deal with multiple stuck faults and certain models for short circuits, but with the FAF technique it is very simple to add new logical fault models of arbitrary multiplicities.

The FAF technique is capable of generating all useful tests and fault coverage information for all faults in a single pass through the circuit, provided sufficient memory is available. For large circuits it may be necessary to split faults up into groups and do several passes, but this is still much better than one pass for each fault.

Certain circuits, such as adders, cause severe problems despite being relatively small. Such circuits realise boolean functions that are not easily expressed in sum of products form.

If a block of logic were specified to perform a particular boolean function, or to implement a particular truth table, FAF generation could be achieved directly from the specification rather than from the circuit description. This would be much simpler computationally but would be restricted to injecting faults onto primary inputs and primary outputs of the block.

It may be possible to take advantage of hierarchy for larger circuits. Commonly used sub-blocks could be evaluated and their output FAFs stored for re-use (both in the circuit in which they originally occur and in others). This may improve FAF generation time but would not alleviate memory consumption problems, because ultimately the output FAFs of a circuit are the same, no matter how they are derived.

The primary reason for working forwards through the circuit when deriving the FAFs was to avoid evaluating any part of the circuit more than once. This led to the decision to use sum of products form (stored as linked lists of terms stored bitwise) throughout the generation process. This was to allow simple memory management and the use of bitwise operations when manipulating FAFs (which it was thought would be very fast). However, the results obtained both in terms of CPU usage and memory usage have been worse than anticipated. It may, therefore be worth generating the FAFs by an alternative method (such as used originally by Miller) and comparing the results with the current method. Use of the single fault assumption and the effect this has on FAF representation and manipulation could be extended to any such alternative method.

8.10 Future Work

On going work on this project will look into;

1) improving the reduction rules for FAF generation to implement all the Zissos reduction rules.

2) using a tree search working backwards from the primary outputs to generate the FAFs as originally done by Miller, but incorporating new manipulation and reduction rules based on the single fault assumption. This method could then be compared with the current technique.

3) implementing post processors to extract various test sets (go/no-go, location etc) and identify untestable faults, directly from the FAFs thus eliminating the need to generate the test functions.

4) making use of a hierarchical circuit description.

8.11 References

COOK S.A., "The Complexity of Theorem Proving Procedures", Proceedings of the Third Annual ACM Symposium of Theory of Computing 1971, pp 151-158.

MILLER P.J. and Taylor G.E., "Automatic Test Generation for VLSI", Proceedings of the International Conference on Circuits and Computers 1982, pp 452-455.

POAGE J.F., "Derivation of Optimum Test Sets to detect Faults in Combinational Circuits", Proc. Symp. Mathematical Theory of Automata, Polytechnic Institute of Brooklyn, pp 483-528, 1963.

ZISSOS D., "Problems and Solutions in Logic Design", Oxford University Press, 1976.

Chapter 9
Macro-Test: A VLSI Testable-Design Technique

F. Beenker and R. G. Bennetts

9.1 *Introduction*

Designing testable VLSI devices presents a continuous challenge to VLSI full-custom designers. Design-for-testability (DFT) has emerged as an integral part of the design process, but the integration can only be achieved if the right tools are in place. In this chapter we discuss the concepts of macro-testability and present the underlying tools to allow designers of VLSI devices to implement the testability structures required by macro-testability. These tools are now in use within Philips and the chapter concludes with comment on the practical application of such techniques.

9.2 *Testability*

The word "testability" has been defined in the following way (Bennetts, 1984):

"An electronic circuit is testable if test patterns can be generated, evaluated and applied to meet predefined fault detection, fault location and pattern application objectives within given cost and time constraints. Objectives are measured in terms of fault cover, diagnostic accuracy, pattern run time, etc."

Within the life cycle of an electronic product, a number of tests are applied, e.g. device test, loaded board test, system test, field-maintenance test (Claasen *et al*, 1989). Test engineers create test programs for each stage of testing to meet such objectives and within real-time constraints. If the test objectives cannot be met, then the circuit is said to be untestable. The consequence of an untestable circuit is that there is an increased risk of a faulty unit-under-test (UUT) being passed by the tester (a test "escape"), or that the identification of the precise cause of failure cannot be made resulting in an inability to carry out a zero defect program. This in turn means that the first-pass yield cannot be improved with a consequent impact on profitability.

Escapes can also impact profitability simply because the product released to the customer contains defects. The customer is the final tester and if the product does not come up to expectation, then it is likely that the customer will not buy from such a supplier in the future. In other words, the supplier is seen to deliver poor quality products. Testability, therefore is seen to be a fundamental characteristic of a product, and the starting point is with basic building blocks - the VLSI devices. But, testability cannot be added as an afterthought once the design is complete. Testability must be considered as an integral part of the design process and must be included within the design specification. As designs become more complex, and as designers make more use of proven sub-circuits (macros), then the testability techniques also become more complex. Figure 1 shows the typical design flow of a VLSI device designer.

Designers work from the highest level of the design ("design a multi-processor array") down through the hierarchy ("use an array of processing elements plus some ROM and RAM") to the lowest level ("Construct the processing element from local RAM plus arithmetic circuits and busses"). They then work back up through the hierarchy, creating higher-level macros from established low-level macros and creating the data-path and control infrastructure between the macros.

Test engineers tend to work from the lowest levels up to the highest levels. That is, they work out what stimuli to apply to the low-level design entity and what response to expect. Then they determine how to apply the stimuli to the inputs of the low-level design entity and also how to set up transfer paths from the outputs of the low-level design entity, working from the accessible circuit inputs and outputs. In the case of a VLSI device, the accessible inputs and outputs are the pins of the device, and in a truly hierarchical design, establishing the input and output paths for the stimuli and response can be an extremely difficult if not impossible task without some thought given to the test infrastructure through the design. This is the philosophy of macro-test (Beenker *et al*, 1990).

182 Macro-Test: a VLSI testable-design technique

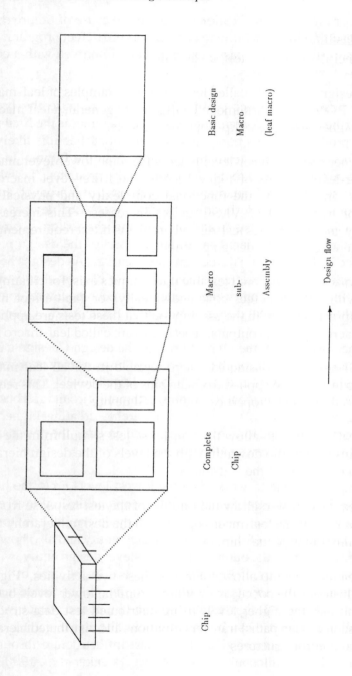

Figure 1 A typical design flow

9.3 Principles of Macro-Test

The principles of macro-test are as follows:

Basic design entities are called leaf-macros. Examples of leaf-macros are; RAM, ROM, PLA, Arithmetic Units, etc. In general, a leaf macro is at a level higher than gates, flip-flops, or registers.

A designer creates higher-level macros from the lower level macros. The higher-level macros of today become the lower-level macros of tomorrow. In this way, the functional complexity and physical size (number of transistors) of VLSI devices increases. This increase in complexity and size creates test-related problems if test requirements are not considered during the design process.

A leaf-macro is considered testable if techniques exist for creating test patterns with known defect coverage (ideally) or fault-effect model coverage (practically) with the assumption that these tests are applied to the leaf-macro inputs and outputs. These tests are called leaf-macro tests.

A VLSI device is considered macro-testable if, for all leaf macros, leaf-macro tests can be applied from the pins of the device. This requirement infers three DFT properties of the VLSI device:

1) features exist to allow the transfer of test stimuli from the input pins of the chip down through the levels of the design hierarchy to the inputs of the leaf-macro;

2) features exist to allow the transfer of the test response from the outputs of the leaf-macro up through the design hierarchy to the output pins of the chip;

3) features exist to allow control of these two activities. Figure 2 illustrates the concept of working from the lower levels back up through the higher levels using distributed test data structures (such as scan paths) transfer conditions and distributed hierarchical control structures.

184 Macro-Test: a VLSI testable-design technique

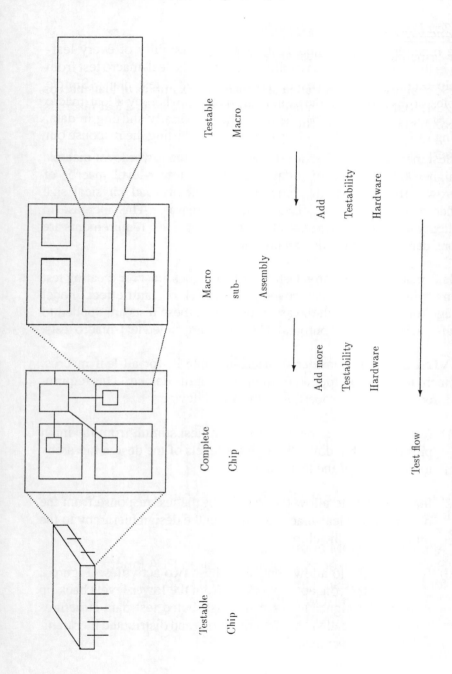

Figure 2 The macro-test principle

9.4 A Test Plan: The Macro Access Protocol

The main criteria in Marco-Test are the accessibility of every leaf-macro in the circuit and the possibility to execute the leaf-macro test from the chip pinning. This access protocol is called a test plan. Simple examples of test plans are the scanning protocols whereby a leaf-macro is observable and controllable via scan-chains: serially shifting in data, applying this data to the leaf-macro and serially shifting the response out for observation, see Figure 3.

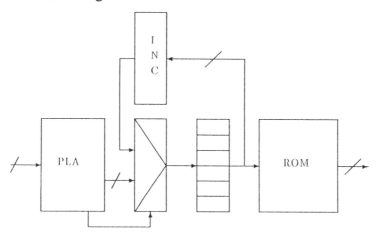

Figure 3 A collection of macros

An outline test plan for the ROM, avoiding scan in Figure 3 could be as follows:

Apply initial ROM address via input;
Switch Multiplexer by applying required PLA input;
for ROM_address = initial to maximum do
    begin   read ROM output
                 increment address
    end

Similar test plans can be created for the other macros and the glue logic.

One of the criteria for testability is to perform all tasks in a cost effective way. This might result in rejection of the simple "scan" test plans due to the resulting long test time and additional silicon real estate. The best possibility to provide this flexibility is to allow a wide range of test plans. The reasoning behind this is that the more knowledge is used about the circuit, the less additional testability hardware has to be added to the circuit to guarantee leaf-macro access.

The Sphinx system, developed at the Philips Research Laboratories to support test requirements of the Piramid VLSI Design System (Beenker *et al*, 1990 and Woudsma et al, 1990), is based on the macro-test philosophy and uses the concept of test plan generation. Test plan generation starts from the test application protocol of a leaf-macro and generates the test application protocol of this leaf-macro described from the chip pinning. For this purpose, data transfer paths of the surrounding cells are used. The current functionality is that these transfer paths are transparent modes of existing cells or of extra DFT hardware (such as scannable flip-flops).

The test plan generation is done on the basis of three types of information: the circuit netlist (macro interconnect), initial test plans for the leaf-macros in the netlist (how to test a leaf-macro as a stand alone unit) and the transfer descriptions of the cells in the netlist. Note that the leaf-macro contents are not required. Only possible transfer information should be provided, which is a subset of the total leaf-macro functionality. Test plans are generated either automatically or under guidance of the user by manipulating the test plan manually (partly or completely).

The test plan contents consists of a control set, a stimulus set and a response set. The control set defines the conditional values for the data flow from the chip pinning to the leaf-macro pinning and reverse (e.g. it defines the condition for scanning through the scan-chain). The stimulus set defines the circuit pins where test data has to be applied for testing a leaf-macro. The response set defines the circuit pins where test data has to be observed from a leaf-macro. The timing relationship between all terminals is specified. Note that the test plans are test data independent. The merge of test plans and leaf-macros test data is done after the test plan generation process. This assembly process requires the leaf-macro test data and the test plans.

9.5 DFT Rules for Macro-Testability

9.5.1 Leaf-macro testability rules

The most commonly used leaf-macro test methods are scan-test and built-in self-test. Various implementations exist of both methods. We only will describe the scan-test rules as used at the Philips Research Laboratories.

Scan design is an attempt to reduce the complexity of the test-generation problem for logic circuits containing global feedback and stored-state elements. The global feedback causes a dependency of the future state on the present state of the device. This dependency causes all the problems in test generation. The primary inputs and outputs are the only device pins over which direct access is possible.

The scan design technique provides a solution to this problem by providing extra access to the stored state elements and as such reducing the circuit complexity. Additional testability hardware is provided with the circuit, such that

1) The stored-state elements can be tested in isolation from the rest of the circuit;

2) The outputs of the combinational next-state decoder circuit that drive into the stored-state elements can be observed directly;

3) The inputs of the next-state decoder and output combinational circuits, driven by the stored-state elements, can be controlled directly.

The method by which this is achieved is to establish a scan-path through the stored-state elements as shown in Figure 4.

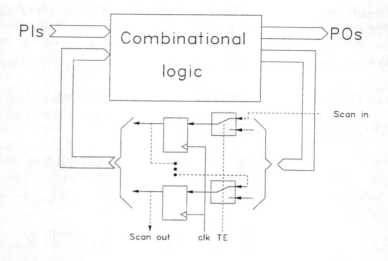

Figure 4 Principle of scan-path

Effectively, each stored state element is now preceded by a multiplexer under the control of a common Test Enable signal. When Test Enable is off, the multiplexers connect the outputs from the combinational logic through to the input sides of the stored-state elements, i.e. the circuit works in its normal mode. When Test Enable is on, the stored-state elements are reconfigured into an isolated serial-in, serial-out shift register. In the scan mode therefore, the stored-state elements can be preset to any particular set of values by placing the values in sequence on the scan data input and clocking the shift register.

The benefits of scan-design are clear. Rather than test the circuit as an entity, the addition of the scan paths allows each major segment (stored-state elements and combinational logic sub-circuits) to be tested separately. The only test generation problem is to generate tests from the combinational segment. This is a solved problem (PODEM, FAN,.....)..(Bennetts, 1984)

Various implementations exist of scan. The following set of rules is used from the Piramid VLSI Design system.

1) All stored-state elements are Master-Slave D-type flip-flops.

2) Every D-type is preceded by a two-way multiplexer. One input of the multiplexer serves as a serial test input. The other input serves as the normal functional input.

3) The multiplexer is controlled by a static control signal.

4) The D-type is clocked by two non-overlapping clocks.

5) The clocks of the D-types must be directly controllable.

6) Clock primary inputs can only be connected to D-type clock inputs or to a clock-generator which generates the two non-overlapping clocks. They cannot be connected to D-type data inputs, either directly or through the combinational logic circuit.

7) All D-types are connected to form one or multiple scan-path registers with a scan-in primary input, scan out primary output and accessible clocks.

8) No loops are allowed in a scan-path.

9) No logic is allowed between the D-type output and the test data input of the next D-type in the scan-chain, except for the last D-type to a primary output. Between the last D-type output and the primary output a two-way multiplexer may exist. This allows to multiplex a serial-out to be multiplexed with a primary output, saving extra circuit pins.

9.5.2 DFT rules for test data access

The access protocol to every macro is described in the test plan. The test plan contains a description of the data access and the corresponding control for guaranteeing the access. The Sphinx system makes use of various forms of data access:

Transfer paths: A cell in a design may have a so-called transfer property. This means that under a certain condition any value on a given set of input ports is copied or inverted (transferred) to a given set of output ports. A transfer property is a subset of the function of a cell. It might be used to transfer stimuli from the circuit pins through the cell to a macro under test and to transfer responses from a macro under test through the cell to the circuit pins.

Scan-chain: A special case of a cell having transfer properties is a scannable register cell. Various forms of scannable cells exist in the Piramid VLSI Design System. Basic data-scan chain cells perform the standard functions "scan" and apply/observe. Bus control block scan cells are able to drive data onto a bus line, receive data from a bus line and to scan data. Instruction scan cells have the additional functionality above data scan cells that they can hold data for a number of clock-cycles (the so-called hold-mode). Test interface scan-cells (Beenker *et al*, 1990) have a transparent mode of operation (similar to boundary-scan cells). Finally, self-test scan cells are used in self-testable leaf-macros to perform the functions scan, normal operation and self-test operation (Dekker *et al*, 1988).

The Sphinx system provides tools to route these scan-cells automatically. Criteria for routing are minimisation for test-time under user-defined constraints (Oostdijk *et al*, 1991).

Direct access: The most straightforward approach to provide access to a macro is via a direct access from the circuit pins to the macro ports. This access is normally provided via multiplexing (again, a cell having a straightforward transfer property).

The notion of transfer property allows to scan-access to be regarded as being a special case of data access to a macro.

9.5.3 DFT rules for test control access

Transport of test data is only possible under certain conditions. Control signals are needed to start self-tests, to take care of scanning in, applying and scanning out test patterns through scan-chains, to enable transfer properties, etc. Hence, not only a test data access has to be provided. Also test control access has to be available.

Several possibilities exist to provide the test control access. In most cases, this access is provided via additional circuit pins. An example is the Test Enable pin required for serial scan data access. However, providing control signals through bond pads may lead to a large number of extra IC pins. To avoid the use of too many extra pins, all test control signals might be generated by an on-chip Test Control Block (TCB). In this manner, testability becomes a real design issue: separate design of a data path with corresponding design synthesis activities. Through the TCB opportunity, the IC designer is able to trade off bond pads and control wiring with silicon area needed for the TCB.

9.6 The Test Control Block Concept

9.6.1 Introduction

The Sphinx system offers the opportunity of generating all test control signals by an on-chip TCB. The TCB is implemented as a finite state machine that produces all control signals needed. To control the TCB itself at least one pin is need. Through the option within Sphinx the IC designer is able to trade off bonding pads and control wiring with silicon area need for the TCB. In realistic designs about one percent of the silicon area of the complete IC is taken up by the TCB (Marinissen and Dekker, 1991).

Recently, proposals concerning testability and on-chip test control hardware have gained significant attention (IEEE Standard, 1990 and Avra, 1987). All proposals are based on the use of a standard test bus combined with a TCB. However, these proposals work with fixed size and functionality of the TCB or give a straight implementation of the required test control procedures, while no hardware minimisation is performed. Within Sphinx a flexible TCB concept is implemented.

9.6.2 TCBs in hierarchical designs

In designs with more that one hierarchical level there are two alternatives for handling the TCBs. The first alternative provides a possibility of using a macro with its TCB as a black box in the design. This results in a herarchical structure of TCBs, which reflects the hierarchical structure of the design. This structure is called the TCB-tree, see Figure 2. The leaves of this TCB-tree are the finite state machines used to generate the control signals for the low-level-macro tests. The task of the higher-level TCBs is to select the underlying TCBs sequentially. The communication between two TCBs of different levels is done via two wires. A second alternative is to extract the TCBs out of the leaves of the TCB-tree and assemble them into one high-level TCB, see Figure 5.

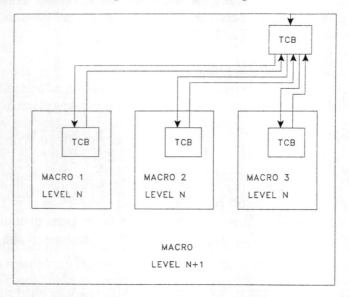

Figure 5 Example of hierarchical structure of TCBs

Considerations while making the choice between a hierarchical structure of TCBs and an extracted and assembled TCB involve chip pinning and silicon area to be used to the TCB and routing of its control busses. A hierarchical structure will demand less area for routing of the control busses, because the width of the control words is determined locally. On the other hand, low-level TCBs are very often similar, so that especially at low-level in the hierarchy an extracted and assembled TCB is expected to save TCB area, because many TCBs can be merged into one.

Up till now we only have experience with the non-hierarchical alternative. TCBs of various functional building blocks were extracted and assembled into one TCB for the entire IC. Conceptually it is also possible to implement the first alternative. It can be expected that with the increase of the complexity and number of hierarchical levels of the designs, hierarchical structures of TCBs will be implemented too, if only to prevent the busses from becoming too wide.

9.7 Sphinx Software Overview

9.7.1 The Sphinx tools

Sphinx has essentially two parts. The first part is a *"testability synthesis"* part whereby an interaction with the design takes place in order to produce, or synthesise, a testable design. During this phase the decisions and corresponding actions are taken to come up with an economically viable testability implementation. The second part is a *"test synthesis"* part whereby all the data manipulation is taking place. The essential point is that the interactive design process is separated from a batch oriented test data manipulation process.

As explained, all flexibility in the system comes from the notion of a test plan and the ability to generate test plans and to implement required testability hardware. Starting with an *initial test plan*, the Test Plan Generator is able to generate a test plan for every macro under guidance of the user by a so-called *guiding test plan*. A guiding test plan indicates the conditional setting which is required for the test data flow. The Test Plan Generator takes care of the administrative role to extract the corresponding data path in the design and to produce the so-called *generated test plan*. In case the test plan generator indicates a problem in access to a certain leaf-macro, action has to be taken by the user to create this access. Besides the possibilities to indicate the test plan manually (the user defined test plan) or to add more transfer information to the design, he has the choice to modify the design. In this process he is supported by Sphinx tools.

User defined test plans can be given as an input and transfer information can be provided to the system. The insertion of testability hardware is supported by tools such as the Test Control Block Generator, the Scan Chain Router and the Boundary-Scan Insertor.

After this process, a stamp should be given to the design to indicate a successful pass of the first phase of Sphinx. This stamp is a combination of a successfully generated test plan for every macro and passing the Design for Testability Rule Checker. The IC design is checked for compliance with a set of basic testing rules. The checking is performed on four items:

1) Scan-chain connectivity (to bond pads)

2) Existency of required elements in the netlist

3) Consistency of netlist connections

4) Check on dangling terminals

For each macro, test patterns defined at the macro terminals are automatically created by a range of Test Pattern Generation tools based on a set of macro-specific fault models. Tools are available for random logic macros (both combinatorial and scannable), PLAs, RAM, ROM and on-chip boundary-scan logic. The interfaces to and from these tools are standardised. This provides an open framework to interface with other TPG tools.

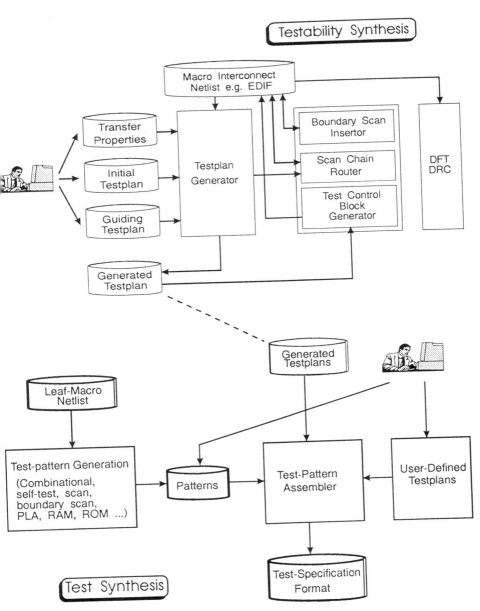

Figure 6 Sphinx overview

The Test Pattern Assembler takes care of the administrative task of merging test plans and test data resulting into a Test Specification Format (Sebesta *et al*, 1990), which is independent of the target tester or target software-simulator, but which can be easily translated into either of these environments. The Test Pattern Assembler works on the basis of either the generated or the user-defined test plans and the leaf-macro test data. This test data either comes from the TPG tools or is defined by the user. In this manner, the Test Pattern Assembler is also capable of helping in design verification. The user indicates the access protocol for controlling or observing nodes in the circuit and the Test Pattern Assembler creates the corresponding Test Specification Format. This approach is similar to the concept as used in the DN10000 project as described in (Bulent, 1988).

9.7.2 The Sphinx user interface

The user interface has been built using the X-Windows system, and more specifically the OSF/Motif toolkit. This toolkit is used to create the textual parts of the interface, menus, text inputs etc.

9.8 *Sphinx Application Examples*

Added Hardware	Extra Area
22 Bus Control Blocks 1 Test Control Block 141 Scannable Flipflops Wiring	0.8% 1.0% 1.5% 2.5%
Total	5.8%
Scanpaths Number of Vectors Clockcycles	3 21853 2690466

Table 1 Initial test-performance data

Sphinx has been used and is currently in use for a number of large industrial VLSI devices. The circuit which has been published is an error corrector (ERCO) (Beenker *et al*, 1990, Oostdijk *et al*, 1991 and Marinissen and Dekker, 1991) for a compact disc player, containing about 150k transistors. This design was created with the Piramid silicon compiler and is a good example to show the trade-off analysis within the Sphinx system.

At the time of the design, the Sphinx system did not contain the full scan-chain routing tool-set. Scan-chains were routed manually and the result was that the test data volume was too large (above the limit of 512k test data volume per tester pin). For this reason, it was decided to make the 16k embedded SRAM self-testable and to minimise on the test data volume for one of the non-selftestable embedded SRAMs. For this reason, a minimal test algorithm was chosen, with a resulting degradation in fault coverage. The initial test-performance data is given in Table 1. The results obtained after insertion of the SRAM self-test and another choice of test algorithm for the embedded SRAM are given in Table 2.

Added Hardware	Extra Area
22 Bus Control Blocks	0.8%
1 Test Control Block	1.0%
141 Scannable Flipflops	1.5%
1 RAM Self - test	5.6%
Wiring	3.0%
Total	11.9%
Scanpaths	4
Number of Vectors	1681
Clockcycles (w/o BIST)	210125
Clockcycles (BIST)	428768
Total Clock Cycles	638893

Table 2 Original test-performance data

The number of test vectors which is indicated in the tables is the number of vectors to test the leaf-macros of the ERCO as stand alone units. For obtaining the total number of required clock cycles we have to multiply the number of vectors per leaf-macro by the number of clock cycles required for test data transportation.

The self-test of the RAM in the ERCO is optional. We compared what the result would be if the RAMS were accessible via scan and the complete set of test patterns were applied. With the help of the scan routing tools in Sphinx, the scan-chains were re-routed and the total number of test data vectors was within the stated 512k limit. Hence, the overhead of the RAM self-test could be removed with a reduction in test time! The results are indicated in Table 3.

Added Hardware	Extra Area
22 Bus Control Blocks	0.8%
1 Test Control Block	1.0%
141 Scannable Flipflops	1.5%
Wiring	3.0%
Total	6.3%
Scanpaths	4
Number of Vectors	21853
Clockcycles	490087

Table 3 Optimised test-performance data

Other designs, of complexity of roughly 500k transistors are currently being dealt with with the Sphinx system. Continuously, a trade-off is taking place between additional silicon real-estate, pinning, test-time and test generation time. Typically, the silicon real estate figures are 5% with four additional pins (test control and some scan-out pins which are not multiplexed with functional pins). The designs all have application in consumer electronic products, whereby this trade-off analysis is of great importance. Detailed results will be published in subsequent papers.

9.9 Acknowledgement

The authors acknowledge the contribution of the members of the Test Working Group at the Philips Research Laboratories.

This work was partly funded by the European Commission through ESPRIT2 project EVEREST-2318.

9.10 References

AVRA, Paper 41.2, Proc. IEEE International Test Conference, 1987, pp964-971.

BEENKER F., DEKKER R., STANS R., VAN DER STAR M., "Implementing Macro Test in Silicon Compiler Design", IEEE Design and Test of Computers, April, 1990.

BENNETTS R.G., "Design of testable logic circuits", Addison Wesley, 1984.

BULENT I.D., "Using scan technology for debug and diagnostics in a workstation environment", Proceedings IEEE International Test Conference, Washington, 1988, pp 976-986.

CLAASEN T., BEENKER F., JAMIESON J., BENNETTS R.G., "New directions in electronics test philosophy, strategy and tools", Proc. 1st European Test Conference, pp5-13, 1989.

DEKKER R., BEENKER F., and THIJSSEN L., "A Realistic Self-test Machine for Static Random Access Memories", Proceedings IEEE International Test Conference, Washington, 1988 pp 353-361.

IEEE STANDARD, 1149.1, IEEE Standard Test Access Port and Boundary-Scan Architecture, IEEE, New York, 1990.

MARINISSEN E-J., DEKKER R., "Minimisation of Test Control Blocks", Proceedings 2nd European Test Conference, Munich, 1991.

OOSTDIJK S., BEENKER F., THIJSSEN L., "A Model for Test-time Reduction of Scan-testable Circuits", Proceedings 2nd European Test Conference, Munich, 1991, pp 243-252.

SEBESTA W., VERHELST B., WAHL M., "Development of a New Standard for Test", Proceedings of EDIF World, New Orleans and IEEE International Test Conference, Washington D.C., 1990.

WOUDSMA R., BEENKER F., VAN MEERBERGEN J., NIESSEN C., "Piramid: an Architecture-driven Silicon Compiler for Complex DSP Applications", Proceedings ISCAS Conference, New Orleans, May, 1990.

Chapter 10
An Expert Systems Approach to Analogue VLSI Layout Synthesis

M.F. Chowdhury and R.E. Massara

10.1 Introduction and Background

The production of computer-aided design tools and environments for analogue VLSI circuit layout has proved very much more difficult than in the digital domain. The principal reason for this is that analogue circuit design involves the accommodation of many, often conflicting, practical constraints (Allen, 1986). Typical of these constraints are the optimum placement of circuit components, layout of large transistors, routing of interconnection channels noting the importance of avoiding signal crosstalk, and the minimisation of parasitics which might affect circuit performance (Haskard and May, 1988, Kimble et al, 1985 and Serhan, 1985). In addition, the design of analogue circuits requires precision modelling of individual components and extensive simulation to verify the desired performance (Rijmenants, 1988). The problems associated with analogue VLSI layout techniques are reasonably well understood by experienced analogue circuit designers and it is widely accepted that these problems cannot be solved purely algorithmically. To produce a sophisticated analogue VLSI layout design tool which will enable these demanding multiple-constraint problems to be solved effectively, requires the efficient incorporation of design expertise as an integral part of the automatic analogue VLSI layout design tool. The purpose of this work is essentially to make a contribution towards meeting these demands, and to provide a greater understanding of typical expert layout design rules and how they should be incorporated within the Expert Analogue Layout System (EALS).

Typical examples of the entities that arise in the integrated circuit (IC) realization of analogue systems are input and output pads, amplifiers, filters, comparators, voltage- and current-controlled oscillators, phase-locked loops, A-D and D-A converters, and sensors (Haskard and May, 1988). Analogue systems are essentially constructed by combination of such circuit elements, which are usually regarded as functional cells or modules within the system. Substantial research effort is being devoted to developing these and similar cells so that design can be carried out at the system level without having to be concerned with the intricacies of low-level analogue circuit design and physical layout on chip (Rijmenants, 1988, Kayal *et al*, 1988 and Tronteli *et al*, 1988). To date, there are very few commercially-available automatic software tools available for the design of analogue VLSI layouts. Amongst these software tools are ILAC (Rijmenants, 1988), SALIM (Kayal *et al*, 1988), and CHIPAIDE (Makris *et al*, 1990). ILAC has a number of attractive features, however, the use of the system is limited because only algorithmic principles are employed. On average the system is said (Rijmenants, 1988) to produce layout which is 25% larger than a hand-crafted layout. On the other hand, SALIM is a system which makes use of both algorithmic and knowledge-based approaches to produce an automatic layout of an analogue VLSI circuit. The algorithmic rules used in SALIM are said (Kayal *et al*, 1988) to be borrowed from the digital domain. The layout tasks in SALIM are considered at the physical silicon-layout level, which is usually time consuming and computationally intensive. For placement SALIM uses a force-directed approach (FDA) (Quinn and Breuer, 1979) which only seeks to minimise the wire length. One major problem with SALIM is that the system checks for analogue constraints only when the routing is complete. If there are any violations, a fresh start has to be made.

The approach adopted in EALS is in many respects quite different from the techniques described above. Firstly, EALS is based on both algorithmic and knowledge-based approaches. The layout generation is not heavily dependent on pre-layout simulation and this is particularly facilitated by the initial planning mechanism (Hammond, 1989 and Wilkins, 1988). The most novel feature of EALS is its unique ability to learn and generate the rules from the input netlist description in order to deduce whether a presynthesised layout information exists in the knowledge database. The layout design tasks are considered in a hierarchical manner and the system is designed such that it is able to fully exploit this hierarchical organization. The main aim of our design concept is to minimise the need for user-defined information or user interaction, while maximising the flexibility of the layout design produced.

The main function of an AI-based planning scheme is to generate a goal solution (an analogue circuit layout in this case) which corresponds to the expectation of an experienced analogue layout designer. To achieve this, various subgoals have to be satisfied before the final solution is reached. These subgoals are categorised into essentially two groups of tasks. The first group of tasks is concerned with the *placement* of components and the second is concerned with the *routing* of the nets connecting the placed components.

Section 10.2 describes the overall features of the EALS expert layout system, and gives the parsing procedure used within it in detail. Section 10.3 describes the implementation of the system and presents the heuristics which play a key part in the way that the system achieves its results. The treatment of layer types is described, and examples illustrating the operation of the system are given.

10.2 Description of the System

A block diagram of the EALS expert layout system is given in Figure 1. The input to the system takes the form of a netlist together with knowledge concerning transistor aspect ratios, and matching information concerning the components or groups of components. The initial input description of the circuit can also be provided with expert knowledge concerning interconnection constraints such as sensitive nets (for example, some nets will be particularly subject to signal crosstalk problems) and maximum allowable interconnect parasitics (Lin and Gajski, 1987). The netlist information is treated as the input database to the system and, where necessary, some parts of the database may be dynamically modified during the execution of the process.

The layout of a VLSI system typically involves five primary tasks:

1) placement of the functional blocks or primitive transistor cells;

2) defining the orientation of the placed components

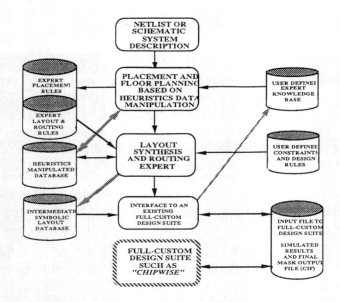

Figure 1 Analogue VLSI expert layout system

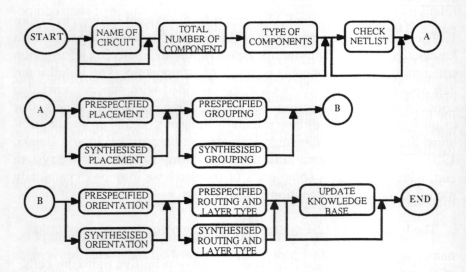

Figure 2 Overview of the planning scheme

3) grouping the components in common wells;

4) routing the interconnects between the functional blocks or the primitive cells

5) defining the type of layer to be used for each interconnect.

A simplified block diagram of the expert planning scheme is shown in Figure 2. The circuits are represented in terms of hierarchical functional modules. To enable the system to produce a layout for general purpose application, the system is equipped with knowledge of how each functional module is to be laid out, and what constraints are to be imposed to achieve the functionality of the overall system. In contrast to the standard-cell approach (Pletersek et al, 1985), where the layout of each functional module is rigidly defined and cannot be altered, here functional modules can be freely altered if it is found to be necessary. The alteration of the layout architecture of a functional module is permissible because the modules are merely intended to provide a guideline to the expert layout system. The functional modules therefore provide information such as the relative location of the components, preferred orientation of each component and the interconnectivity constraints from one component to another.

To synthesise the layout, the first action is to check whether a presynthesied layout can be found which matches the requirements of the input circuit description. To determine if a presynthesised layout is available in the library, the system checks the following information: type of circuit, number of components, type of components, the interconnections list (netlist).

A pictorial representation of the system designated for carrying out this procedure is given in Figure 3.

The initial checking process comprises two phases; in the first phase name, number, and transistor type data are verified and a selection is made of the solutions which most closely match these data. The second phase is concerned with the detailed comparison of the netlist information supplied in the original input circuit description and the netlists selected during the first phase. If the pattern of the selected nets matches exactly that of the original input netlist, the system keeps a record of all the nets which have been tested and repeats the checking process until all the nets connected to each component are successfully tested. The unique feature of the system is that, to perform the parsing operation, the system dynamically generates the necessary program code directly from the information supplied in the knowledge database in order to check the connectivity pattern of each selected component. A diagrammatic representation of the detailed parsing tasks is given in Figure 4.

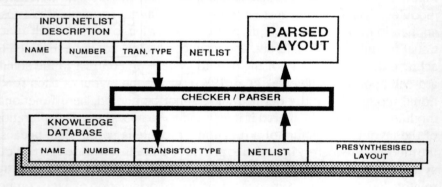

Figure 3   Basic parsing mechanism

The interesting point to note here is that because the system has the ability to generate its own code in PROLOG format (Bratko, 1986) directly from the netlist description, it is not normally necessary for the user or the programmer to exclusively define the rules for various modules, as these rules can be generated automatically and stored away for later use. The task of the parsing system is essentially to establish a correspondence between the input circuit description and one or more equivalent circuit descriptions defined in the knowledge database. The database is organised in a highly modular fashion where the data structure of the original input circuit description is virtually identical to the data structure of the circuit description stored in the system's database.

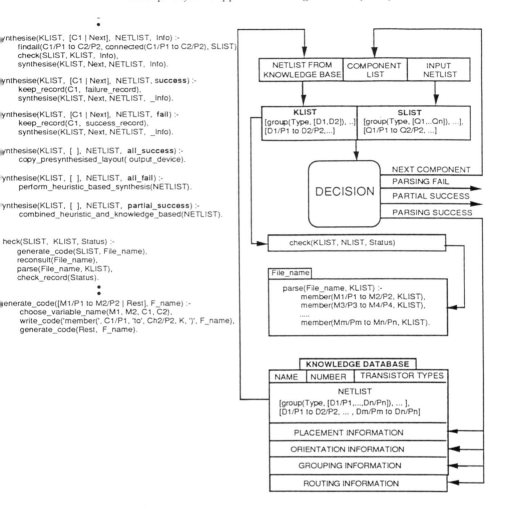

Figure 4 Parsing tasks in detail

It should be noted that the approach adopted here to matching the netlist description is analogous to the process of template matching. One important concept is that the exact ordering sequence of the rules is not important. The idea is that as each net is arbitrarily selected, the system will attempt to satisfy the key features which uniquely define the class or the type of circuit. As each rule is satisfied completely, the corresponding nets are removed from the list.

On completion of the initial checking procedure the system may either find a complete solution, a partial solution or no solution.

10.2.1 Complete solution

A complete solution is found when the system is able to check successfully all the information supplied in the input circuit description compared with the circuit description of a selected presynthesised layout. If this is the case, the system simply copies the presynthesised layout information directly from the knowledge database.

10.2.2 Partial solution

Partial solutions arise in two instances. The first is when the system finds that, although the initial checking procedure was a success, the layout information provided in the knowledge database does not provide a complete solution to the problem. If a complete solution is not found the system has to generate the "missing" information by executing the appropriate tasks. For example, the database may only provide the information concerning the *placement* and the orientation of the components without providing the functional grouping and *routing* information. In this case the system will have to generate these data independently, using the primitive rules which are specified in the system's database. Once this information is generated, the user can instruct the system to keep a permanent record of the synthesised layout.

The second instance is when the initial checking procedure has failed, but there is an indication that some of the presynthesised layout can still be used if the problem is decomposed at the next lower level within the hierarchical planning scheme. At this level, each component is grouped in terms of a *functional module*. Usually the type of functional module formed by any individual component is readily specified with the input netlist description of the circuit, so the system does not normally have to search the connectivity information exhaustively to determine to which functional group a particular component belongs.

Generally when a partial solution is encountered the system will still have to generate the *placement, orientation, functional grouping, routing* and *layer-type information*. The principle used to find the goal solution is highly systematic, in that the approach used for one level of the hierarchy is equally applicable to any other level in the hierarchy.

10.2.3 No solution found

If the input circuit is not defined in a hierarchical manner in terms of a set of functional modules, but instead is defined at the transistor level, then a different approach is applied. In this approach, the system starts the layout synthesis process by first placing the component selected as having a highest *weighting value* at a location predetermined by the algorithmic heuristic evaluation process. Subsequent sections will detail exactly how this placement-criticality weight is computed. If the placed component belongs to a prespecified functional module then the other components forming that particular module are heuristically selected and placed as dictated by the same algorithmic process.

The internal database created for the final layout is translated into a format compatible with the GEC "STICKS" symbolic layout representation (Cope *et al*, 1986). The symbolic layout file is then accessed from "CHIPWISE", a full-custom IC design suite from which the physical layout is generated (Walczowski *et al*, 1989). The "CHIPWISE" suite can be used interactively to modify the design if needed, and the final layout can be simulated with extracted parasitic capacitances. At this point, of course, any other full-custom design package available to the designer can be used.

10.3 Implementation of the General Concept

10.3.1 General placement concept

Most expert systems require some form of inference engine which combines algorithmically-computed cost values with information supplied in the knowledge-base to enable the system to make a quantitative or qualitative decision to solve the problem (Carley and Rutenbar, 1988). The goal of the expert system here is to use this inference engine to drive the system to determine the relative position of each placed component on a two-dimensional virtual grid plan. To ensure that the circuit components are optimally placed (i.e., would correspond to the expectation of an expert layout designer), the placement tasks are conducted in two stages as shown in Figure 5. Firstly, a heuristically-arranged plan of the circuit components is built to determine a selection sequence criterion for each component. Once the selection sequence is established, the system attempts to generate the placement using a rule-based expert systems technique (Buchanan and Shortliffe, 1984). Secondly, if the system is unable to realise the complete goal solution using the rule-based technique, the empirical and epistomological heuristic methodology (Pearl, 1984) is used to deduce the "best" placement for each heuristically-chosen component. Figure 6 gives an example of how these rules are represented.

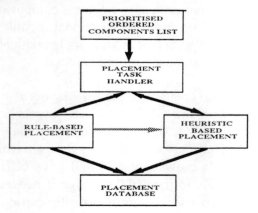

Figure 5 Placement options

An expert systems approach to analogue VLSI layout synthesis 211

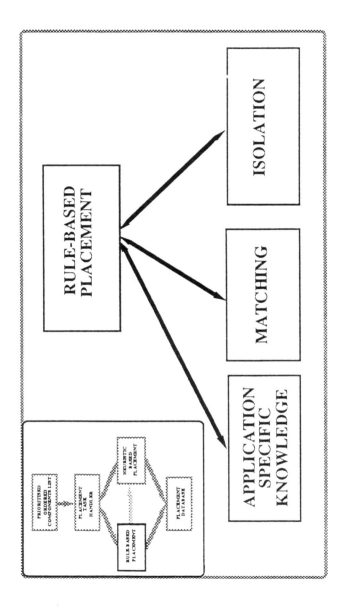

Figure 6 Rule-based placement options

10.3.1.1 Application-specific knowledge

The idea underlying the approach introduced here is to exploit the use of previously optimised placements or layout architectures by simply-attempting to match the netlist description by searching the library of prespecified knowledge. This option is essentially provided to allow the system to generate highly application-specific placements - that is, those not normally solvable by heuristic algorithmic means. Figure 7 gives a pictorial representation of this approach, featuring an op amp as an example of a typical analogue circuit module.

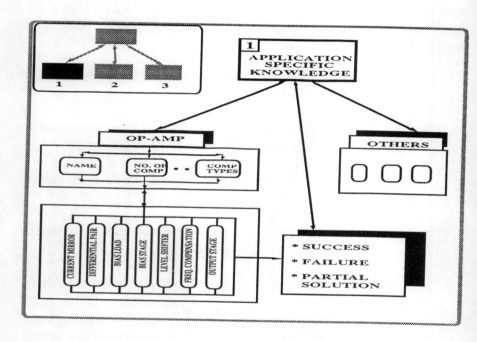

Figure 7 Application-specific representation

10.3.1.2 Matching

One of the most critical design constraints, which directly regulates the performance of an analogue circuit, is the accuracy with which the components/functional blocks are matched. It is virtually impossible to match components exactly in every practically desirable respect. This is purely because of significantly large variations in the silicon fabrication process. However, the effects of the mismatched components can be substantially improved by observing the following design guidelines (Gregorian and Temes, 1986 and Kimble et al, 1985). The guidelines have been compiled from experienced analogue and digital layout designers and they are used to form a basis for the establishment of the expert design rules:

1) Place the components that need to be matched adjacent or as near to each other as possible.

2) Where possible, place matched components in common wells; especially for the most critically-matched components.

3) Avoid mixing digital and analogue circuits within functional blocks.

4) It is generally good practice to place all similar functional blocks as near to each other as permissible.

5) Avoid placing components with comparatively large power dissipation directly adjacent to matched components. If this is not possible, either try to distribute the high-power-dissipating components evenly around the matched components, or use appropriate precautionary measures (Gregorian and Temes, 1986).

10.3.1.3 Isolation

A second important factor with significant impact on the performance of an analogue VLSI circuit is the proper isolation of the various components or groups of components. The main purpose of isolating components is to minimise interference between different sections of the circuit.

Many forms of interference are encountered by layout designers; some are more significant than others. The exact behavioural nature of a source of interference is often very difficult to model, and designers normally have to resort to some form of approximation technique (Haskard and May, 1989 and Williams, 1984). Listed below are the most common forms of interference encountered:

1) Temperature mismatches.

2) Spurious feedback and parasitic effects.

3) Loading effects and impedance mismatches.

4) Crosstalk caused by large voltage or current swings.

5) Unexpected interferences, particularly when digital and analogue circuit elements are mixed.

10.3.2 Heuristic decision evaluation for placement

One of the major problem of developing a rule-based expert system is that there are always situations when a solution can be found in more than one way (Wilkins, 1988). When these conflicting situations are encountered, the problem is to find a single optimal solution. The question is how does the system determine which is the best option, and what criterion does it use to justify its choice? For these types of situations, the only way the problem can be resolved is when some form of heuristic evaluation process is employed to facilitate the decision-making process (Cho and Kyung, 1988).

The first task the system has to perform is to determine how each component is to be placed. To serve this purpose, it is appropriate to categorise each circuit component as a list of prioritised options. Each component is assigned a weighting value. The *weighting value* is computed as a function of the number of interconnects and the type of constraint(s) imposed on the component.

The equation to evaluate the initial weighting value is given by (Chowdhury and Massara, 1989):

$$W_i = \sum N_{ij} \tag{1}$$

where $j \neq i$, and N_{ij} is the number of *direct* connections from component i to component j.

Once this measure of component interconnectivity is computed, the exact component-selection sequence is determined by making a quantified comparison of the constraints imposed on each component as a function of the weighting value W_i. The quantification process is essentially an analysis of the functional property of the component i. For example, if the component i forms part of a current mirror, a constraint on i may be imposed specifying that i has to be matched with another component k. Since the matching of components is frequently critical for the correct functionality of an analogue circuit, the weighting value of component i (W_i) is adjusted such that the system would select this component for placement at a very early stage in the placement sequence. The final quantified weighting values are thus computed and collected into a list WL[1, ..., m] in descending order to signify the relative priority of each component in the circuit. The organised list thus enables the system to determine the choice between the *highest priority component* (HPC) and the *lowest priority component* (LPC). It is normally found that the HPC is the device having the greatest number of interconnects and the most critical constraints. On the other hand, the device with the least number of interconnects and fewer constraints is usually classified as the LPC.

The system now treats the HPC as the "seed" component. The seed component is used to determine which is the next best component to be selected, and where this component should be placed. It should be noted that there are situations in which more than one HPC component may result, particularly for situations when there are, for example, several identical current mirrors in a given circuit. If there are two or more HPC components, the constraint information for each component is checked.

This involves summing the weighting values of the subsequent components W_j which are directly connected to the HPC components.

$$W_i{'} = W_i + \sum_j W_j \qquad (2)$$

Where j is the index number of a component that is connected directly to component i.

If there is still more than one instance when the new weighting values are found to be equal, then the constraint factors between the components i and j are taken into consideration. In this case the components associated with the most critical constraints are considered first. If a decision cannot be reached at this stage then an arbitrary selection is made. The first selected component is treated as the seed component since the components which are subsequently selected are directly influenced by it.

Once the seed component has been placed, the subsequent unplaced components are either selected automatically as a result of the previous computation stage, or by computing a heuristics estimation of the attraction between the unplaced components with respect to the component which has already placed (Chowdhury and Massara, 1990). The unplaced component which produces the highest attraction value is then selected for placement.

The computation of the relationship between the placed component j and the unplaced component i, can be expressed by :

$$R_{i,j} = n * (W_j + W_i) \qquad (3).$$

where W_j is the weighting value of placed component j, W_i is the weighting value of unplaced component i, and n is the number of connections from component i to component j.

Once again a situation may arise when there are two or more instances at which $R_{i,j}$ is found to be equal. When this situation is encountered, the weighting values of each unplaced component i are evaluated for each available location immediately adjacent to the placed component j. Equation (4) describes the formula for the evaluation of the weighting values at each given location x, y.

$$L_{x,y} = K * \sum_j \left[\left(\frac{W_i}{(C_i + 1)} + \frac{W_j}{(C_j + 1)} \right) * \frac{N_{i,j}}{Ed_{i,j}} \right] \quad (4)$$

Where

Lx,y is the weighting value at location x, y
Wi is the weighting value of the placed component i
Wj is the weighting value of the unplaced component j connected to component i.
Edi,j is the mean Euclidian distance from component i to component j, i.e.

$$Ed_{i,j} = \sqrt{((x_i - x_j)^2 + (y_i - y_j)^2)}$$

Ni,j is the number of connections between components i and j.
Ci is the cost introduced on component i by component j at location x,y.
Cj is the cost on component j when it is placed at location x,y.
K is an adjustment factor in multiples of 10s - it determines the accuracy of the computed weighting value (Lx,y).

The cost factors C_i and C_j are calculated only in relation to the supply pads where the supply pads V_{DD} and V_{SS} are placed at the top and bottom of the chip geometry respectively. The values of C_i and C_j are computed such that when the component j is placed at a location x,y, if at this location component j obstructs the supply connection of component i then C_i is set to 1, otherwise C_i is set to 0. Similarly, when component j is placed at that location and if its supply connection is obstructed by component i then C_j is set to 1, otherwise it is set to 0.

Having computed the weighting values at each location, and if there exists a single highest value, then the associated component is selected and placed at this optimum location. If, however, there is still more than one instance where the weighting values are equal, the final stage of the placement decision-making phase is executed. Here, the system attempts to make a decision based on the estimation of the any sensitive nets and whether these nets could be connected directly with a minimum number of connection bends once a particular location of the placed component is selected. On completion of this final stage, if still no decision can be reached because of more than one instance when the final stage is found to be satisfied, then an arbitrary choice is made.

10.3.3 Component orientation

The general organization of the planner to facilitate the orientation of components is similar to the plan developed for the placement of components. That is, in the first instance, the system always attempts to determine the component orientation using the rule-based technique and only if this fails will the system resort to solving the problem using a heuristic evaluation process. The rule-based technique involves use of the basic information such as the type of surrounding components, the connectivity between components and the relative position of the placed components. In order to determine the orientation of each component, the system treats this information as a prerequisite for a predefined target solution. The target solution is only found when all the prerequisites are satisfied. The plan shown in Figure 8 is essentially based on a *case-based planning scheme* (Hammond, 1989). This plan is valid for any level of abstraction, whether at transistor- or systems-level.

When producing the layout at the transistor level there are essentially two different orientations; vertical and horizontal (Cope *et al*, 1986). Vertical orientation is denoted by '1' and horizontal orientation by '0'. Normally, for the purpose of simplicity and ease of compaction, the initial orientation is selected directly from the aspect ratio of the transistors such that the size of the components in at least one direction is kept constant. For example, if there are transistors T1 and T2 such that T1 has dimensions L=10 and W=40, and T2 has dimensions L=60 and W=10, assuming the size of the components are fixed in the x-direction then orientation =1 is selected for T1 and orientation = 0 is selected for T2. This is illustrated in Figure 9.

Using this simple orientation convention, it may not always be possible to produce an optimally-compacted layout or properly-oriented components, particularly when it is desirable to match the components in a prespecified manner. For example, it is normally desirable to design the layout of a differential input stage as shown in Figure 9(a). But, because the aspect ratios of the transistors T3 and T4 are such that L > W, the orientation code=0 is selected, whereas for the transistors T1 and T2, the orientation code=1 is selected since L > W. The alternative layout for this specific situation is shown in Figure 9(b). These problems therefore cannot be solved using the above simple rules and for most practical examples it is imperative to use knowledge-based techniques to solve the problem.

The initial orientation of the transistors does not provide information concerning position of the source, gate and drain pins. Once the initial transistor orientation is selected, there are basically four different pin combinations that are available; the task of the system is to select the 'best' combination for a given placement. This is achieved by evaluating the weighting values for each pin combination and selecting the combination which gives rise to the highest weighting value. The equation for the weighting value evaluation is given by:[3]

$$WP_i, pin = \sum_j \frac{W_j}{E_d*(N_0 + 1)} \quad (5)$$

where E_d is the Euclidian distance from $x_i - y_j$ to $x_i - y_j$, and N_0 is the number of obstacles between components i and j. The combined weighting value of the component for a given orientation is given by:

$$WT_i, orient = \sum_{pin}^{Max} WP_i, pin \quad (6)$$

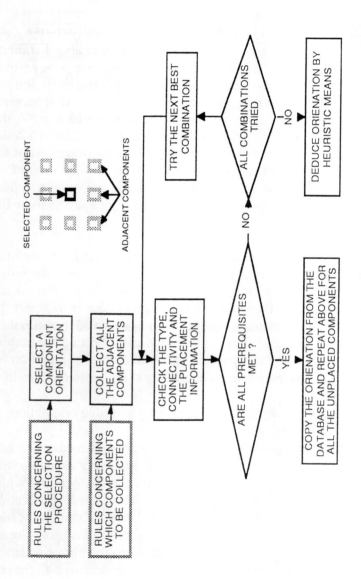

Figure 8 Rule-based component orientation plan

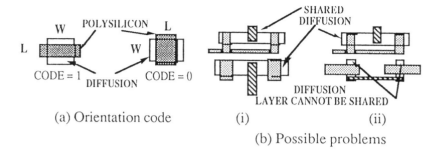

(a) Orientation code (i) (ii)
(b) Possible problems

Figure 9 Problem of simple aspect-ratio-based component orientation

The methodology for determining the orientation of each component involves two basic tasks :

1) Selection of the components according to their relative priorities.

2) Evaluation of weighting values at every possible orientation.

For each pin, we compute a weighting value giving a measure of how favourable a particular choice of pin location is. For every available valid pin location, the weighting values are calculated and the locations which give the highest weighting values are chosen. There are fundamentally two options available to the system in making a decision about the location of each pin. In the first option, the system makes the decision directly from the calculated weighting values derived from the above equation. The second option involves making a combined assessment of all the pins associated with a component i; the combination which produces the highest estimated weighting value is then chosen.

Let :

LP$_{i,k}$[g] be a list of g elements holding the weighting values kth pin of a component i.

g be an integer number, 1, ..., G, which acts as a pointer to each available location.

G be the maximum number of pin locations available adjacent to the components.

Option 1:

e.g: choose g, such that LP$_{i,k}$[g] is a maximum. Use g to find the corresponding location, x$_g$, y$_g$.

1) Select the elements of the list, LP$_{i,k}$[g], for which the weighting values are an absolute maximum.

2) Compare each selected location in turn; if there exists a unique location which satisfies a given goal solution then the corresponding location is chosen.

3) If more than one predefined goal is satisfied, and if a priority order can be found, then the location with the highest priority option is chosen.

4) If a priority order is not found then option 2 is executed. This option effectively enables the result to be found using the combination of all the associated pins of the chosen component by taking the average highest value.

Option 2:

1) Select each pin combination in turn and compute weighting values (C$_u$) at every available location, using equation (7), and insert the computed values in a list.

2) Arrange the list in descending order and choose the first element of the pin combination from the list produced in step (i) which has the highest weighting value - if alternative options exist, they are ignored.

The combined weighting value (C_u) of all the pins is given by :

$$C_u = \sum_{k=1}^{M} WP_{i,k} \qquad (7)$$

where $WP_{i,k}$ is as given in equation (5), M is the maximum number of pins associated with component i, u is a pointer to each combination set defined in the database.

The component orientation option is chosen when C_u is a maximum. On completion of this stage, a database with the relative positional information is created and it is used by both the component grouping task handler and the layout routing synthesis processes.

10.3.4 Grouping of components

Our aim is, of course, to produce analogue VLSI layout that is optimal in every respect. Grouping a set of components into a common well is an important factor that should be considered. The function of the grouping task-handler is to ensure that components of the same type are either grouped together and placed in the same well, or isolated by placing them in separate wells. The grouping operation therefore requires extensive use of expert knowledge database information. Figure 10 gives a representation of the grouping task handler. A detailed description of how the grouping rules are represented is explained with the aid of a case study example later in this chapter.

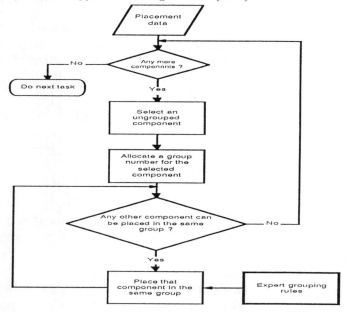

Figure 10 Component grouping task handler

Once their orientations have been established, the components are grouped together in their respective wells. The grouping of the components is dependent on the relative location of the placed components, their types, and the specific rules concerning how the components are to be grouped. There are two options that are available when performing the grouping tasks. The first option is where the system attempts to group the components using the primitive rule-based techniques. If the system is unable to find a solution using the rule-based technique, then the problem is solved using a heuristic algorithm. In the algorithmic approach the components are grouped depending on the type and location of the adjacent component. Only components of the same type are grouped together in the same well. In some situations, the algorithmic approach will not be able to satisfy the specific analogue layout design constraints. For example, it may not be desirable to group a pair of components which would normally be grouped and placed in the same well. A typical example of this type is when two matched component

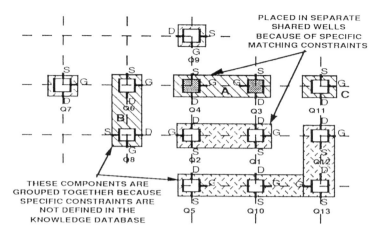

Figure 11 An illustrative representation of how components are grouped

are grouped together and it is found that a third unmatched component could be placed in the same group. If the third component is placed in the same well as the matched components, this could seriously affect the overall performance of the circuit. Once again, problems of these types cannot be solved purely algorithmically and, as a result, the system has been equipped with a primitive set of rules which enable it to decide how the components are to be grouped. Figure 11 gives an illustrative representation of how the components are grouped for an op amp example. It should be noted that the component placed in group C is not merged with group A to ensure that the matching requirements of Q3 and Q4 are not violated (Tronteli et al, 1987 and Kayal et al, 1988). In the same way, group B is not grouped with group A. Similar reasoning also applies to all the other components.

10.3.5 Routing between interconnects

In the placement stage we have essentially built a paradigm for the expert system. This paradigm is used by the system to make an intelligent decision about the layout task in order to find the "best" solution to the given problem.

In solving the routing problem, the first task is to build a plan that will solve the problem. As a part of this initial plan, the system first transforms the placement database to give sufficient grid-point spacing to ensure that interconnect routes are always found. For this reason we have found that it is quite acceptable to assume a spacing of twice the number of total interconnects between each placed component. This assumption is quite legitimate, since these are virtual grid points and they do not have any major effect on the actual layout; they only affect the overall processing speed during the physical layout generation phase (Cope et al, 1986). It is possible to employ an optimisation technique to find the optimum spacing between each group of placed components. For the present system, optimisation techniques have not been used, and we have used the simple spacing method suggested above. It should also be noted that allowing the maximum number of grid points spacing between the placed components enables the system to find routes with a minimum number of overlaps between two different types of layers. This subsequently allows the system to minimise the parasitic capacitances. It also allows the routing planner to find a shorter path where appropriate.

To establish the execution sequence of the case-based plan that is developed for the layout synthesis, a strategic decision criterion was formulated as explained below. It is worth mentioning that considering the layout task in a modular fashion simplifies our task; we have given the system the ability to make use of existing and well-established algorithms to find a solution to the problem. For example, instead of using the Manhattan line search routing method, we could use other types of routers (maze router, lee router etc.) In formulating the strategic decision criterion, we note that routing selection sequence plays an important role in the architecture of the final layout. In view of this, we have once again used a quantified decision criterion. The quantification process will effectively give a measure of the relative importance of the component in relation to the functional behaviour of the overall circuit. The computation of the quantification value can be summarised by four basic tasks:

1) Select the source and destination item from the netlist.

2) Determine relative distance between the source and destination components.

3) Make use of any prespecified constraints given in the knowledge-base.

4) Attempt to satisfy these constraints by making sure that the routing resistive and capacitive parasitics are kept at a minimum.

The above information is formulated as:

$$P_{s-d} = \frac{H_s + H_d}{D_{s-d}} + M_f \tag{8}$$

where

P_{s-d} is the weighting value of source and destination points.

H_s is the heuristic estimate of the source component.

H_d is the heuristic estimate of the destination component.

D_{s-d} is the Euclidian distance between source and destination component.

M_f is the matching constraint factor, given by:

$$M_f = \frac{Degree\ of\ importance\ indicated\ in\ the\ knowledge-base}{Number\ of\ obstacles\ between\ s\ and\ d}$$

If there are multiple paths between the source and destination points, the choice is made by taking routing parasitics into consideration. The parasitic estimation is a function of the length of the path and the number intersections along it.

The P_{s-d} values are computed for all the source and destination points from the netlist. These are then stored in a list in descending order so that the first item of the list is considered to have the highest priority and the last item of the list to have least priority. Once the decision criterion is established, layout synthesis is performed by partitioning the netlist in two groups. The first group consists of the single connection list. The second group consist of the multiple connection list. Both are of course arranged in a descending order with appropriate decision factors as computed previously. The single connection sequence list is used first to find the routing path using the case-based plan which is developed for this purpose. The multiple-connection sequence list is then used to find the remaining path. During the multiple-connection path finding sequence, additional plans designed specifically for solving the multiple connection problems are also used. As mentioned previously, we have treated the layout tasks as a route-finding problem. It is therefore important at this stage to consider what our goals are, and how we are going to achieve them.

10.3.5.1 What are the goals?

The first and foremost goal is to exploit the route-finding techniques to not only find the actual routes of all the netlist elements, but also to use the techniques as an integral part of the layout generation. This approach has the benefit of eliminating one of the design steps and it also introduces an extra degree of design flexibility (Dai and Kug, 1987). Simultaneous consideration of the layout and routing task is permissible because they both share a common objective - to minimise parasitics and to keep the layout as compact as possible. We find the method used for parasitic minimisation for the layout generation is similar to the minimisation technique used by the routing system.

When routing analogue VLSI systems, as already noted, one of the prime objectives is to ensure that parasitic effects are minimised. To achieve this, there are sets of design guidelines that need to be followed since there is no known single algorithmic solution to the problem. These guidelines are treated by our system as a set of expert rules, and these rules are the tools that are used to design a case-based planner with the intention of finding a solution to any general analogue VLSI design problem.

10.3.5.2 How are the goals achieved?

The most general rules-of-thumb for minimising circuit parasitics are (Haskard, 1988 and Gregorian and Temes, 1986):

1) Ensure that all the interconnect wires are kept as short as possible.

2) Minimise the number of interconnect crossovers.

We have classified these rules in a set of prioritised groups. The highest priority group consists of all the rules associated with the matching of the circuit elements. When a group of components is to be matched the rule is to try to avoid finding routes through the matched components, other than those associated with the matched components themselves.

The route-finding process has two prime tasks; (a) to find the "best" route and (b) to avoid any routing conflicts. Both of these tasks are carried out by single- and multiple-connection path-finding steps.

10.3.5.3 Single connection path-finding case-based planner

The planner proposed here is based on the above rule-of-thumb and tries to satisfy this general concept whilst observing the general rules of analogue VLSI layout design. An intelligent planner must be constructed so that it knows its ideal target goal. From this target goal, it must at least be able to find the nearest goal solution if it is unable to find the exact one. It must have the ability to remember failures and successes. Both items of information are equally important (Hammond, 1989). It is also desirable for the plan to be constructed using a minimal set of rules, whilst maximising the use of the existing plan (Hammond, 1989, Ho et al, 1985 and Hu and Chen, 1990).

To find the path of a given source and destination point we have adopted a technique similar to Manhattan line search method (Asano, 1988). The general principles for finding the routes are as follows:

1) Treat all the previously-placed components as a list of obstacles. Construct a set of primitive templates.

2) Use these templates to probe between the locations A and B to find the shortest "best" path.

The general structure of the case-plan developed to deal with the task of global route finding is given below. We have used the term "global" here since at this stage conflicts are ignored. The planner is able to produce a list of option paths and order them according to their preferences, so that if needed the system can suggest alternative paths without having to repeat the whole process. It should be noted that the paths evaluated are treated by the system as only a set of suggestions, as the solution cannot be guaranteed until all the conflicts are removed. There are some rare cases when the system may not be able to remove all the conflicts; in these cases the next best alternative will tried.

This strategy uses the plan developed for the first-order conflict eliminator as described above, but in addition it introduces new segments, usually at either one or both ends of the path. Figure 12 illustrates the general concept.

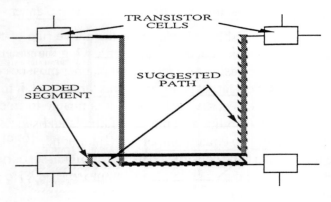

Figure 12   Conflict removed by adding new line segments

10.3.5.4 Multiple connection path (MCP) case-based planner

Our task is to find a path from a point X_1, Y_1 to X_2, Y_2, where X_1, Y_1 represents the position of pin A and X_2, Y_2 denotes the position of pin B. Our precondition is that either pin A, pin B, or both, are already used by other pin connections, and paths for these connections readily exist in the layout database. The MCP problem is therefore to find the "best" path from the paths that use pin A and the paths that use pin B. Preconditioning steps before executing the path finding plan are summarised below.

10.3.5.5 Execution sequence generation

A list of possible source and destination points is compiled and arranged in order of priority.

1) Fetch all segments that use Pin A --> List A,

2) Fetch all segments that use Pin B --> List B,

3) Compare segments of List A with List B --> Produce a list of paths --> SHORT_LIST.

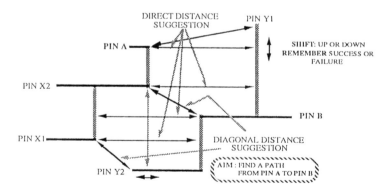

Figure 13   Suggested path options

SHORT_LIST is evaluated as a function of the Euclidean distance from the source and destination points, combined with the number of obstacles and intersects between the two points. Note that this preconditioning stage, where possible, suggests the most direct connection. This is illustrated in Figure 13.

The important thing to note about this routing algorithm is that the routes are found using a template fitting topology. The routing process begins by firstly finding the routes of all the nets which form a direct connection (without any bends) from a given source and target locations. The first nets to be routed are those which do not intersect with other nets. For this situation the path of the route is simply a straight line from the source and target locations. If, on the other hand, a net forms an intersection with any other net(s), the system checks to ensure that these nets do not cause any design rule violations or at least prevent or minimise the violating situation, if it is able to do so effectively and without any additional overheads. When confronted with this type of situation the problem is solved by firstly checking if the intersecting nets do indeed introduce any serious routing violations. If no violation is found, the suggested routing paths are accepted and the system continues with its next sequence of operations. If however, a violating situation is detected, then the system identifies the most sensitive net as the net which needs to be routed with a direct straight line connection. If several equally sensitive nets are found, a selection is made on the basis of the net forming the shortest estimated distance.

Having found the route of all the straight line nets, the next phase of the routing task is executed. Here all the remaining nets, including the nets which failed at the previous stage, are collected, analysed and partitioned into two groups. The first group consists of all the singly-connected nets. The second group consists of all the nets which form multiple or junction connections. The partitioning of the nets is established according to the information supplied in the knowledge base concerning the functional properties of the source and destination components and the primitive rules that are applicable for the given spatial and parametric information.

An expert systems approach to analogue VLSI layout synthesis 233

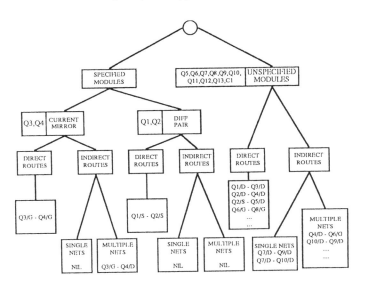

Figure 14 Op amp hierarchical routing tasks example

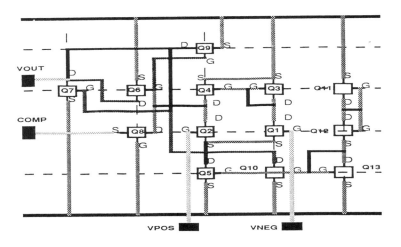

Figure 15 Initially-routed layout without the layer-type information

As a general design strategy, the sequence at which these nets are routed is determined by the sum of the weighting values of the source and target components. The weighting value of each net is calculated from the individual weighting values of the source and target components of the net. The net which gives rise to the largest weighting value is selected. If there is more than one instances when the weighting value is observed to be the largest, then the distance between source and target locations is taken into consideration. The net with the shortest distance is routed first. If there are multiple source and target locations registered as the shortest distance, a decision is made by examining the type of pins to which the net is connected. If the source and target pins are of the same type (e.g. source-to-source, drain-to-drain, etc.), and it is unique, then that net is selected. However, if there are still multiple instances, an arbitrary selection is made. A breakdown of the routing tasks for a typical op amp example is shown in Figure 14

As can be seen from the figure, and as already stated above, the layout is synthesised in a hierarchical and a modular fashion. The first module that was synthesised was the current mirror, followed by the differential pair. Figure 15 shows a typical example of applying this technique.

10.3.6 Layer-type definition

Once all the components have been placed, and the paths for the interconnect routes have been identified, the next task is to determine the type of fabrication layer to be used for the interconnects. The paths for the interconnect routes are defined in the form of chained line-segments. Using the symbolic layout technique it is possible to select any type of fabrication layer for each segment of the interconnect path. It is important to choose the right combination of the layer types in order to gain optimal performance from the circuit. The "correct" use of the layer-type influences the following main factors:

1) Matching the components.

2) Minimising the number of contacts used, thereby reducing spurious parasitic effects.

3) Avoiding unnecessary violation of fabrication layers.

4) Minimising the overall routing parasitics.

10.3.6.1 Rule-based layer-type definition

As usual, the rules are defined in a hierarchical manner with the most application-specific rules placed on the highest level of the hierarchy and the most general rules defined in the lowest level of the hierarchy. The type of fabrication layer used for each segment of the routing path is dependent on the accumulation of spatial information combined with specific user-defined information. The spatial information includes details such as

1) The location of the source and destination points.

2) The type of source and destination components.

3) Crossover violations with other layer(s).

Specific user-defined information includes details such as

1) The type of functional element to which the path segment belongs.

2) Sensitivity requirements.

3) The width of the track itself.

The layout design fabrication technology used for the purpose of this research work has five different layer types, five types of contacts and two types of wells. Table 1 gives a list of all the possible layers, contacts and wells.

Type of layers	Type of contacts	Type of wells
P-Diffusion	cut	P-well
N-Diffusion	via	N-well
Polysilicon	p-tap	---
Metal1	n-tap	---
Metal2	butting	---

Table 1 Fabrication technology and the basic layout elements

Figure 16 gives an example for each of the layer types. The figure illustrates the importance of using the correct fabrication layers for each interconnect segment. From these examples it is evident that the layer-type definition tasks are as straightforward as initially envisaged. The exact type of fabrication layer used for each segment is dependent on the extent to which the system is able to manipulate the spatial and the user-defined information to reach a decision which is acceptable by both the system and, more importantly, by an expert layout designer. The system designed to deal with this task is shown in Figure 17.

As can be seen from the figure, there are two possible ways that the system can extract the layer-type information. (1) the system attempts to solve the problem using a rule-based approach. In the event of failing to find a solution this way, (2) the system provides an algorithmic solution to the problem, always guaranteeing a solution.

Once a conflict-free route has been found, the system determines the type of layer that needs to be used for each line segment. Unlike standard routing problems where the designer is only concerned with the interconnection between various modules with a prespecified number of layers, here the system not only needs to consider the routing aspects but in addition it has to consider the component-level layout synthesis aspects. The implication of this is that an added complication is involved when defining the fabrication layers of each interconnection path. Exactly which layer is chosen for a particular net is highly dependent on the type

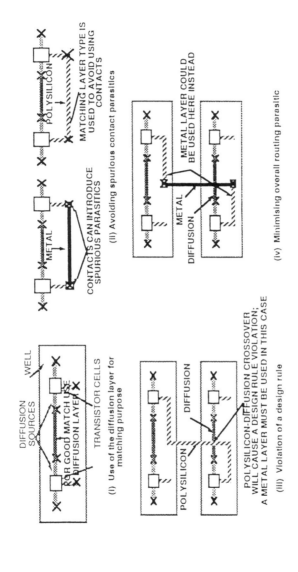

Figure 16 Layer-type definition examples

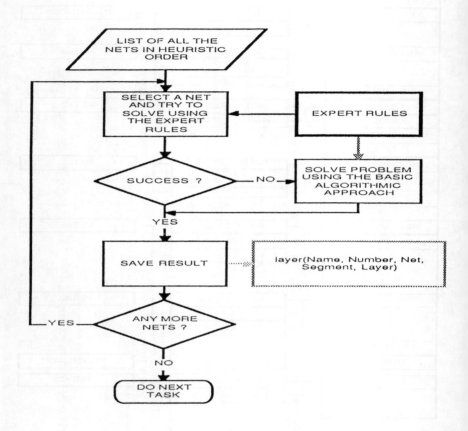

Figure 17   Layer-type definition task handler

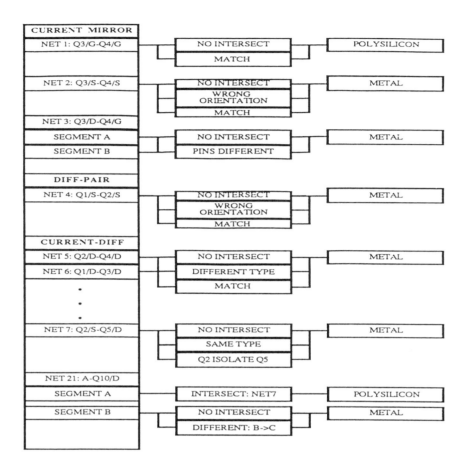

Figure 18 Layer-type definition plan created for an op amp

of source and destination components, type of source and destination pins, the shape of the interconnection path, and the type of layers at the intersection points. An illustration of how the layer types are specified for a typical op amp example is given in Figure 18.

Note that the gate pin of component Q3 in Figure 18 is directly connected to the gate pin of component Q4 by a polysilicon layer. This is a justifiable decision because both Q3 and Q4 are of the same type, the locations of Q3 and Q4 are such that a straight-line connection can be established, and both the source and destination pins are of the same type. For the source pins of the two components, the connection is formed by a metal layer in this case instead of a diffusion layer. This is because, although there is a direct connection from Q3 to Q4, the orientation of the components does not allow the transistors to be constructed on the same diffusion layer; as a result, the components had to be connected by a metal layer.

Once the layout information is generated, the system can be instructed to keep a record of the synthesised information in the library of standard modules. Figure 19 shows the detailed physical layouts of some typical analogue VLSI circuit examples generated by EALS.

10.4 Comments and Conclusions

The Expert Analogue Layout System (EALS) described in this chapter has proved to be a very flexible system. Its unique feature is that most of the physical geometrical properties are, from the viewpoint of the user, kept at an abstract level. This was made possible particularly by using a symbolic layout design topology. The quality of a solution produced by EALS is essentially dictated by two main factors (1) the type of expert rules included in the knowledge database, and (2) the type of symbolic layout topology selected. The symbolic layout topology is important because it directly determines how the actual physical layout is generated. The STICKS symbolic layout topology used by EALS does not allow the inclusion of adequate user-derived information to enable the system to produce an optimally compact layout of the sort that would be produced by expert hand-crafting using the same placement configuration. A better

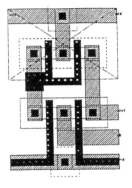

(a) simple differential input stage

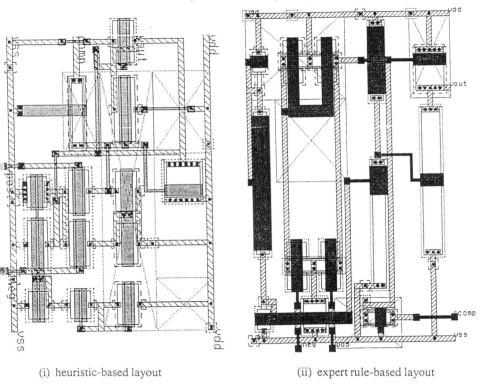

(i) heuristic-based layout   (ii) expert rule-based layout

(b) a two-stage operational amplifier

Figure 19   Physical layouts of some typical analogue curcuits

symbolic layout topology would significantly improve the compactability of the final physical layout. The system is supported with a primitive set of expert rules which are adequate for most analogue VLSI layout applications. Inclusion of expert layout configuration rules is very easy with EALS, since the user only has to supply the circuit description in the form of a typical netlist. The netlist is currently entered in textual form, but a suitable interface would allow it to be directly entered using an interactive symbolic layout graphics editor.

10.5 References

ALLEN P.E., "A Tutorial - Computer Aided Design of Analog Integrated Circuits", IEEE Custom Integrated Circuit Conference (CICC), 1986.

ASANO T., "Generalized Manhattan Path Algorithm with Applications", IEEE Transactions on CAD, Vol. 7, No. 7, July 1988, pp 797-804.

BRATKO I., "Prolog for Artificial Intelligence", Addison - Wesley, 1986.

BUCHANAN B.G and SHORTLIFFE E., "Rule-based Expert System: the MYCIN experiment of the Stanford Heuristic Programming Project", The Addison-Wesley series in artificial intelligence, 1984.

CARLEY L.R. and RUTENBAR R.A., "How to automate analog IC designs", IEEE Spectrum, August 1988, pp 26-30.

CHO H.G. and KYUNG C.M., "A Heuristic Standard Cell Placement Algorithm Using Constrained Multistage Graph Model", IEEE Transactions on Computer-Aided Design, Vol.7, No 11, Nov. 1988, pp. 1205-1214.

CHOWDHURY M.F. and MASSARA R.E., "Knowledge-based Analogue VLSI Layout Synthesis", IEE Colloquium, 6th Nov. 1989.

CHOWDHURY M.F. and MASSARA R.E., "An Expert System for General Purpose Analogue Layout Synthesis", Proc. 33rd Midwest Symposium on Circuits and Systems, 1990.

COPE M.C., REID I.P., PAGE K.P. and JONES M.J., "STICKS - A Symbolic Layout System for VLSI Design", Silicon Design Conference, 1986.

DAI W.M. and KUG E.S., "Simultaneous Floor Planning an Global Routing for Hierarchical Building-Block Layout", IEEE Transaction on CAD, Vol. CAD-6, No. 5, Sept., 1987, pp 828-837.

GREGORIAN R and TEMES G.C., "Analog MOS Integrated Circuits for Signal Processing", John Wiley and Sons, Inc., 1986.

HAMMOND K.J., "Case-based Planning - Viewing as a memory task", Academic Press, Inc. 1989.

HASKARD M.R. and MAY I.C., "Analog VLSI Design - nMOS and CMOS", Prentice Hall, 1988.

HO W.P.C., YUN D.Y.Y. and HU Y.H., "Planning Strategies for Switchbox Routing", Proc. ICCAD, 1985, pp. 463-467.

HU Y.H. and CHEN S.J., "GM_Plan: A Gate Matrix Layout Algorithm Based on Artificial Intelligence Planning Techniques", IEEE Trans CAD, Vol. 9, No. 8, Aug. 1990, pp. 836-845.

KAYAL M., PIGUET S., DECLERCQ M. and HOCHET B., "SALIM : A Layout Generation Tool For Analog ICs", IEEE, CICC, 1988, pp 7.5.1-7.5.4.

KIMBLE C.D., DUNLOP A.E., GROSS G.F., HEIN V.L., LUONG M.Y., STERN K.J. and SWANSON E.J., "Autorouted Analog VLSI", IEEE CICC, 1985, pp 72-78.

LIN Y.L.S. and GAJSKI D.D., "LES: A Layout Expert System", Proc., 24th Design Automation Conference, 1987, pp. 672-678.

MAKRIS C.A. et al, "CHIPAIDE: A New Approach to Analogue Integrated Circuit Design", IEE Colloquium on "Analogue VLSI", Digest No. 1990/073, 1990.

PEARL J., "Heuristics: Intelligent Search Strategies for Computer Problem solving", The Addison-Wesley series in artificial intelligence, 1984.

PLETERSEK T., TRONTELJ J.&L., JONES I., SHENTON G. and SUN Y., "Analog LSI Design With CMOS Standard Cells", Standard Cell Application News, Vol. 2, Number 2, Sept. 1, 1985.

QUINN N.R and BREUER M.A., "A Force Directed Component Placement for Printed Circuit Boards", IEEE Trans., CAS Vol. CAS-26, No.6, 1979, pp 377-388.

RIJMENANTS J., SCHWARZ T., LITSIOS J. and ZINSZNER R., "ILAC: An Automatic Layout Tool for Analog CMOS Circuits", IEEE CICC, 1988, pp 7.6.1-7.6.4.

SERHAN, GEORGE I., "Automated Design of Analog LSI", IEEE CICC, 1985, pp 79-82.

TRONTELI J.&L., PLETERSEK T., "Expert System for Automatic Mixed Analogue Digital Layout Compilation", IEEE, CICC, 1987, pp 165-168.

WALCZOWSKI L., WALLER W. and WELLS E., "CHIPWISE User Manual", The ECAD Laboratory, University of Kent, Canterbury, U.K.

WILKINS D.E., "Practical Planning - Extending the Classical AI Planning Paradigm", Morgan Kaufmann Publishers, Inc., 1988.

WILLIAMS B.C., "Qualitative Analysis of MOS Circuits", Artificial Intelligence 24, Elsevier Science Publishers, 1984, pp 281-346.

Chapter 11
Guaranteeing Optimality in a Gridless Router using AI Techniques

M. F. Sharpe and R. J. Mack

11.1 *Introduction*

CAD tools to support full-custom integrated circuit design may either provide a fully-automated "silicon compiler" (Johannsen, 1979) or a more interactive "design assistant". This second approach can provide the designer with greater control whilst providing CAD tools to free him from low-level tasks such as detailed layout. The design flow within such a design assistant suite (Revett, 1985) consists of three major phases: floor planning, symbolic cell design and chip assembly.

Floor planning describes the topological layout of different functional blocks of the integrated circuit and shows the estimated size and topological relationship between blocks. A top-level master floor plan is divided into elements, each of which may be a symbolic cell, a layout cell or another floor plan. By creating floor plans within floor plans, a hierarchy is realised. Floor plans are created using a floor plan editor and each level of the floor plan describes the relative positions and sizes of blocks and defines the relative positions of interconnect between them.

The leaf cells at the bottom of the hierarchy may be realised directly by the designer or incorporated from a cell library. The adoption of a symbolic sticks representation (Taylor, 1984) for the cell layout frees the designer from requiring detailed knowledge of the design rules for the fabrication process; instead effort can concentrate on the development of architectures to realise the required functionality. When the symbolic cell design is complete the sticks topology is translated into a geometrical layout by a compaction process - the sticks are fleshed-out to appropriate widths for the specified fabrication process and are spaced to produce a cell with minimum dimensions commensurate with the design rules.

Chip assembly produces the overall mask layout for the chip by integrating the layout of the individual cells. The assembler places the compacted functional cells using topological information from the floor planner and interconnects them with routing channels as specified by the floor plan. The tracks within the routing channels are required to match the physical spacing of the functional cells in order to minimise wasted silicon. However the terminals of the abutting compacted cells will not always align directly and the assembly algorithm must employ either cell stretching or river routing in order to align the cells with the routing channels and to complete the geometrical floor plan of the parent cell. This process continues hierarchically until the geometrical mask data for the chip is complete. We now present a method for generating channel routing which conforms to the spacing constraints imposed by the abutting functional blocks in order to minimise the white space created during chip assembly.

11.2 Routing Generation

Lee's original work on path interconnection (1961) was based upon a gridded representation of a maze routing problem and since then a vast number of varied routing techniques have been produced - usually based upon the same gridded representation. A symbolic interconnection pattern is produced which defines the interconnection topology but which consists of dimensionless tracks and contacts. Dedicated symbolic channel routers usually adopt a scheme using two orthogonal layers; they can be very efficient and produce excellent results when the cells interconnected by the routing channel conform to the constraints of the grid. In cell-based systems, for instance, the library cells are usually required to have terminal spacings which match the contact to contact spacing of the fabrication process used. However in full-custom design, the inverse constraint applies and the routing channels are required to conform to the physical spacing constraints imposed by the abutting functional cells. Spacing conflicts can arise between the symbolic pattern and the mask layout of these cells due, for example, to contact dimensions being ignored during routing generation; this results in white space being created in the chip floor plan by the assembler which resolves the conflicts by employing either cell-stretching or river-routing (Pinter, 1983).

These effects are demonstrated with a simple four net example, shown in Figure 1, in which nets 3 and 4 have been compacted to minimum

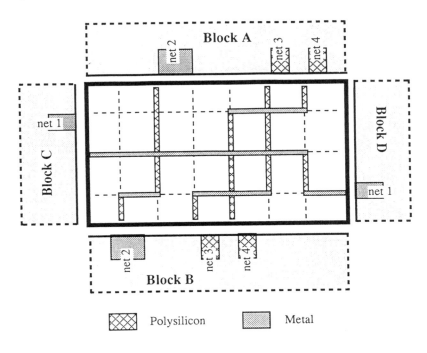

Figure 1 Grid-based routing generation

polysilicon spacing in the functional blocks A and B and net 2 has been specified at twice minimum metal width. The symbolic interconnection pattern produced by a router based upon the grid-based greedy algorithm (Rivest & Fidducia, 1982) is shown in Figure 1 whilst Figure 2 shows the effect of assembling the routing pattern to meet the spacing constraints of blocks A and B. There is an initial problem in that the symbolic router assigns orthogonal directions for the layers; in this case running metal horizontally and polysilicon vertically. This results in layer mismatches on net 2 which require post-processing to perform the necessary layer changes. However the major problem is caused by the finite size of the contacts which are assumed dimensionless by the symbolic router. Assuming the chip assembly spacing algorithm runs left to right, stretches are initially created in block B to align nets 3 and 4 with the routing pattern and then further stretches are needed in block A. Due to the dog-leg paths of these nets more space is opened up in block A due to the "knock-on" effect of contact spacing. Although this is a simple example, we have found in practice that this broadening of white space across routing channels is typical and accumulates across the floor plan creating significant wasted space.

One approach to the problem is to post-process the symbolic routing pattern to meet the spacing constraints of the abutting blocks; techniques include offsetting contacts and performing layer changes. We have found that post-processing only has only limited effect and that a better approach is to generate the routing in a grid-free dimensioned space which accommodates the constraints of the abutting blocks directly. A number of gridless routing techniques have been produced - for example Sato et al, 1980 and Groenveld et al, 1987. Our emphasis has been placed on achieving maximum flexibility and we have developed a single-net gridless Routing Tool to operate within the system architecture shown in Figure 3 (Sharpe & Mack, 1988).

The Task Spotter identifies incomplete nets in the Problem State and passes suggested courses of action to the Planner; this controls the Routing Tool which carries out an individual point-to-point interconnection task using a particular optimisation criterion such as minimum track length or minimum contact-count. The action of the Routing Tool is dependent upon the control of the Planner and varies as the routing continues. The system reads pin positions in absolute dimensions and the design rules of the fabrication process which can include any number of layers. A mask-level routing pattern is produced which can be implemented with a minimum of post-processing.

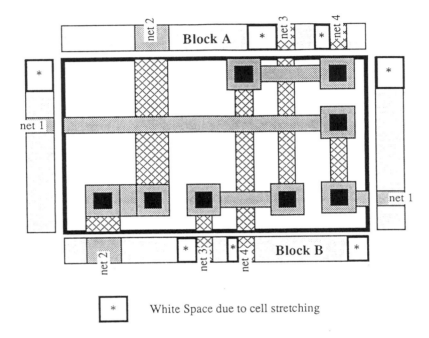

Figure 2 Cell stretching in chip assembly

250 Guaranteeing optimality in a gridless router

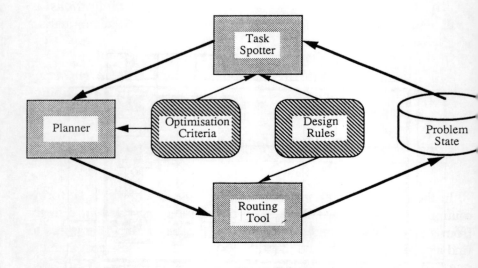

Figure 3 Routing system architecture

The specifications for the routing tool are:

1) It must be *grid-free*. It is concerned with laying out practical geometries on silicon, where conduction tracks cannot be constrained to lie at regular intervals.

2) It must be able to *guarantee connection*. If a path exists, the router must be able to find it.

3) It must take account of fabrication process *design rules*. The routing should satisfy all process design rules without modification.

4) It should be relevant to as *wide a number of applications* as possible, and therefore should not be restricted to one fabrication style.

5) It should be able to carry out routing using a *variable number* of interconnection layers. This gives the tool maximum flexibility and applicability.

6) It should be able to *optimise* the path of any net against user-specified parameters, such as length or contact-count or resistance.

In the following section an algorithm is presented which performs routing in the gridless domain and generates the mask-layout directly from a set of user-specified design rules. The algorithm guarantees to find a route if one exists and that there is no route with less cost than the one found. This is achieved by adopting a technique used in Artificial Intelligence to control the search of an edge-list database. The time complexity of the algorithm is low, and is independent of the resolution of the fabrication process.

11.3 The Gridless Single-Net Routing Tool

11.3.1 Model definition

The routing algorithm which has been developed finds an optimum Manhattan path using variable width tracks across a variable number of layers. In order to introduce the concepts and to maintain clarity in diagrams, the algorithm is introduced in section 11.3.3 by considering a simpler routing model - this is based upon the problem of finding a minimum distance Euclidean path (diagonal distance) across a single interconnection layer using zero-width tracks. The algorithm is based upon a technique used in Artificial Intelligence to find the shortest path through a graph and has been adopted here in order to integrate the production of a graph representing the routing problem with the search of that graph; this approach greatly reduces data generation for routing problems and hence significantly reduces the time taken to find a path. Section 11.3.4 shows how the algorithm is extended to accommodate

Manhattan paths using tracks which are consistent with design rules describing an arbitrary fabrication process.

11.3.2 Representation of the continuous domain

The fundamental difficulty of optimal gridless routing is finding a data structure to represent a continuous routing area. In order to be stored and manipulated by a digital computer, the representation must be discontinuous. The data structure, therefore, must only contain a finite number of positions but must represent a structure with an infinite number of positions. For the purposes of optimal routing, a discontinuous representation can only be considered adequate if the shortest path through the continuous space can always be found by finding the shortest path through the discontinuous data structure representing it.

For the purposes of this gridless routing system, it is assumed that the area available for routing can be bounded by a rectangle. The routing task is defined by a source point and a target point within or on the boundary of this rectangle. All obstacles are rectangular, and their edges are parallel with the edges of the boundary. A graph can be constructed to represent such an area; vertices of the graph are defined at the source point, the target point and at the corners of all obstacles. Two vertices are joined by an edge if a line drawn between the two corners does not cross any obstacles. Edges of the graph can coincide with edges of obstacles, so two adjacent corners of an obstruction may be connected. Each edge is given a weight which is defined by the Euclidean distance between its two end points. This is known as a representation graph. The *representation graph* for a four obstacle problem is shown in Figure 4.

The representation graph that results is discontinuous, and is hence amenable to computer manipulation. To show that this is an adequate representation of the routing space, it is required to demonstrate that the shortest path through this discrete graph will always correspond with the shortest path through the continuous routing space.

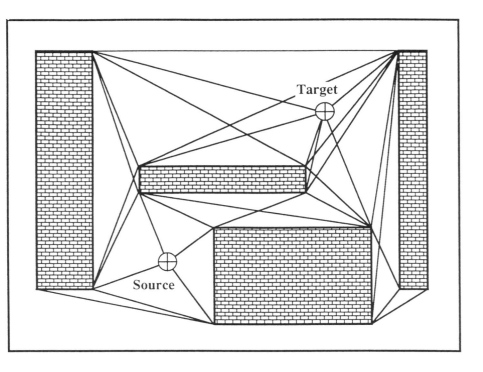

Figure 4 Representation graph for a four obstacle case

A connecting path through the cluttered routing space takes the form of a continuous, piecewise linear track. It can be considered to be a sequence of linear subtracks meeting at corners. No subtrack intersects with an obstacle, although it may coincide with an obstacle edge. Since it is known that subtracks between corners are linear, the connecting path can be completely specified by a list of corners. A subtrack corresponds to an edge of the graph if, and only if, the corners of the subtrack correspond to graph vertices. Hence, a connecting path through the continuous domain corresponds to a path through the graph if, and only if, all corners on the connecting path correspond to vertices of the graph.

By the above argument, it can be seen that a sufficient condition for a connecting path to be represented as a subgraph of the representation graph is that every corner of the connecting path must coincide with a terminal point or a corner of an obstacle in the routing space; this is readily verified by assuming that the shortest path corresponds to that taken by a length of elastic material placed under tension between the terminal points.

Having described the routing space with a discrete representation, it is now required to use this representation to find a shortest path by searching the graph.

11.3.3 A Method for integrated search and graph generation

It has been noted above that the representation graph has the desirable property that it always includes the shortest connection path. However, it has the drawback that complete graph generation is highly computationally intensive - each obstacle has four corners, each of which may be connected to all other corners except itself and the terminal points. For n obstacles, the total number of connection checks which have to be made can be shown to be $8n^2 + 6n$. It is therefore desirable to integrate the generation of the graph with the method used to search it. If it is possible to guarantee a shortest path without searching the whole graph, it is not necessary to generate the portion of the graph that is not searched.

The search algorithm used was the A* algorithm (Hart *et al*, 1968). This is a heuristic search algorithm which is used in Artificial Intelligence to find a shortest path through a discontinuous graph. The A* technique is similar to other search techniques (for example Lee's) in that it maintains a frontier list. At each iteration of the algorithm, the first element of the frontier list is expanded and replaced by its neighbouring cells. If the initial frontier list is the source point, it is known that all possible connecting paths must pass through at least one element of the frontier list at each iteration. In common with other heuristic search algorithms the A* algorithm maintains its frontier list ordered by increasing value of a heuristic function f. However, in the A* algorithm, the heuristic function **f** is defined as the sum of two other functions, **g** and **h**. These represent a known and an unknown quantity in arriving at an estimate.

The function **g** represents the known quantity. It is defined as the total cost of reaching the current node from the starting point. It is calculated from the history of the previous search and is known precisely. Whenever a vertex is expanded, the heuristic cost of moving from the expanded vertex to the successor is measured and added to the g-value of the expanded vertex. This forms the g-value of the successor vertex.

The function **h** represents the unknown quantity. It is an estimate of the cost to move from the current vertex to the target. The function **f**, which is the sum of **g** and **h**, is defined as the cost of moving to the current vertex from the source added to the estimated cost of moving from the current vertex to the target. It is therefore an estimate of the total cost of moving from the source to the target through the current vertex. The form of the estimator function **h** is important. If it can be shown that **h** never gives an over-estimate of the completion cost, it can be guaranteed that there is no path of less cost than the first path found.

Figure 5 demonstrates the A* algorithm solving the Euclidean path problem for which the complete graph was shown in Figure 4. Elements of the frontier list are represented by an expression L:*g*/*f* where L is a reference label and *f* and *g* are the values of **f** and **g** for that vertex. In this case, the cost of moving between vertices is defined as the Euclidean distance between them. This is used to calculate **g** for each vertex. The estimator function **h**, in this case, is defined as the Euclidean distance between the current vertex and the target regardless of the presence of obstacles. The final path length between the current vertex and the target will always be greater than or equal to this. The f-value estimates the total Euclidean path length from source to target given that it passes through the node. Since the **h** function never overestimates, it is known that it is not possible to find a path through a vertex which has a length less than the f-value.

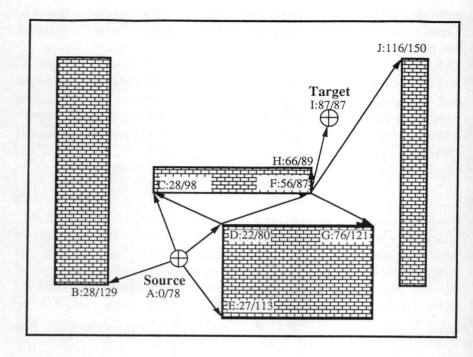

Figure 5 A euclidean path-finding problem using an A*-based technique

A trace of this example is summarised in Table 1. The frontier list is initialised to contain the source vertex. Function g evaluates to 0 for the source vertex and function h evaluates to the Euclidean distance from the source to the target, in this case 78 units. After the first iteration, the best element in the frontier list has an f-value of 80, indicating that the best possible path to the target is now known to be 80 units. This is a tighter constraint than before. As the algorithm proceeds the f-value increases, and the constraint on the path length tightens. After the third iteration, the target is found with a path length shown by the g-value to be 87 units. At this point, the best path that can be achieved by expanding anywhere else in the frontier list is 89 units. It is therefore known that there is no shorter path than the one found. In this case the shortest path has been found in 3 iterations of the algorithm, and only a small proportion of the network has been searched. Since the generating function was controlled by the search process, only a small proportion of the total network was generated. The improvement can be appreciated by comparing Figures 4 and 5.

Iteration	Frontier List
0	[A:0/78]
1	[D:22/80, C:28/98, E:27/113, B:28/129]
2	[F:56/87, C:28/98, E27/113, G:76/121, B:28/129]
3	[I:87/87, H:66/89, C:28/98, E:27/113, G:76/121, B:28/129, J:116/150]

Table 1 Trace of algorithm solving a Euclidean Path Problem of Figure 5

Having established the basic operation of the integrated graph search and generate algorithm, modifications are presented in the following section to constrain the routing to Manhattan paths with tracks of definable widths using multiple layers.

11.3.4 A Practical multi-layer manhattan router

Since tracks in a VLSI device are usually constrained to run horizontally or vertically, the minimum path length of a connection between two points can be found by the Manhattan distance between them. A representation graph for Manhattan path finding can be defined in a similar way as for Euclidean path finding, except that movement is constrained to be either horizontal or vertical. A connection can be made horizontally (or vertically) to the ordinate (or abscissa) of a neighbouring point if the perpendicular from the current location to the ordinate (or abscissa) does not cross an obstacle. A narrow Manhattan-path finding problem and a completely generated graph is shown in Figure 6.

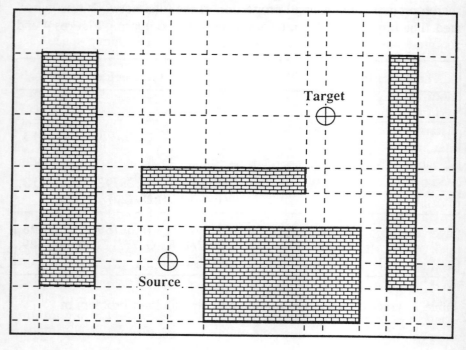

Figure 6 A narrow manhattan path problem with complete graph

In practice, two ordered lists of edge locations are maintained - one for horizontal and one for vertical edges. Dummy positions are added to the lists to represent the source and the target locations. Each legal point in the routing space can be defined by a position in the horizontal and the vertical edge lists. This representation results in a successor function which is far simpler than the Euclidean case. Each vertex can be connected to a maximum of four neighbouring vertices, owing to movement restrictions. The neighbouring vertices can be found by following the pointers in the ordered lists. A connection is allowed if movement between the current point and the next is not blocked by obstacles. This can be deduced simply by examining the current edge list. It can be seen, therefore, that only the local part of the network need be examined. This is a major improvement on the Euclidean successor function, which required a comparison with every other vertex in the routing space.

The heuristic functions required by the A* algorithm have to be modified to take account of the different criterion against which optimisation is to be carried out. The g function is defined as the Manhattan distance from the source to the current point; the h function by the Manhattan distance between the current point and the target ignoring obstacles. This is a consistent underestimator, so the algorithm is guaranteed to return a path with a minimum Manhattan distance.

Having achieved a shortest-path Manhattan algorithm on a single layer, the next requirement is to extend the algorithm to find a shortest path across any number of layers. The method used to achieve this is similar to that employed to extend the Lee algorithm (Rubin, 1974). As well as allowing each vertex to move in four horizontal directions, additional degrees of freedom are added allowing the router to make a connection upwards or downwards. In the case of the Lee algorithm a connection can be made upwards or downwards if the relevant cell is not blocked. This information can be found by addressing a look up table, or similar data structure. In the gridless case, a cell-based addressable data structure can not exist by definition. It is therefore required to maintain a list of open obstacles for each layer. When an edge on another layer is encountered in the edgelist database, the obstacle associated with it is examined. If the edge is a leading edge, the obstacle is added to the list of open obstacles; if it is a trailing edge, it is removed. A change of layer is only allowed if the list of *open obstacles* for the target layer is empty. The technique can be extended to support an arbitrary number of layers, allowing the router to be fully flexible.

The algorithms presented up until this point have found the shortest narrow path subject to certain constraints. In a practical layout, connection tracks are required to have defined widths and separations commensurate with the process design rules. In order for the routing tool to accommodate layer changes, knowledge is required of contact sizes and the corresponding layer overlaps. The Routing Tool operates with appropriate layer dimensions passed from the Planner. For multi-layer routing, each obstacle is entered into the database twice - once expanded by a track separation and once by a contact separation. Edges produced by expanding by the line separation are only relevant to the layer in which the obstacle lies, whereas edges produced by expanding by the contact separation are also relevant in adjacent edges. Each element in the edge-list database has a field indicating which layers it is visible to. Stepping horizontally or vertically through the database is implemented by stepping repeatedly through the edge lists until an edge list which is visible to the current layer is encountered.

11.4 Optimisation Criteria

The algorithm developed so far guarantees to find a shortest Manhattan path route across any number of layers subject to user-defined design rules. Often there are a number of possible connections with the minimum Manhattan distance. In many cases, some of these paths may be preferred over others. For examples, it is often considered beneficial to reduce the number of contacts (Naclerio et al, 1987) or it may be desired to reduce the number of bends in a track (Deutsch, 1985). In order to minimise these criteria, heuristics H_1 to H_n may be combined into single function H such that:

$$H = w_1 (H_1) + \ldots + w_n (H_n)$$

where w_1 to w_n are arbitrary weighting functions. Although good results can be achieved in this way, the quality of the result depends on the weighting functions in a way which is not simply predictable.

The method preferred by the authors is to maintain a number of independent heuristic functions referring to the physical attributes of the interconnect path. These could, for example, be Manhattan distance, the number of bends or the number of contacts. In the A* method, the frontier list is ordered by the first heuristic. Ties are broken by evaluating the next heuristic. This process repeats until either the tie is broken or the algorithm runs out of heuristics, in which case an arbitrary ordering is taken. The resulting path minimises the heuristics in order, and the path reflects the heuristics in a clearly defined way. For example, if the first heuristic were Manhattan distance and the second were the number of contacts, it can be guaranteed that no connecting path could be found with a Manhattan distance less than the one returned by the algorithm, and that there was no path with the same Manhattan distance and fewer contacts.

These heuristics form the optimisation criteria which the Planner passes to the Routing Tool and need not be fixed for the entire channel routing process. The Planner may select specific criteria for special classes of net such as power or clock lines which, for example, may require single layer routing. Figure 7 shows examples where the routing algorithm is called to connect between two different points on the same layer, avoiding a single obstacle. Three optimisation criteria are provided, and the three examples show the path characteristics that can be achieved by specifying a different ordering of criteria in the call to the tool. The three criteria are *distance*, in which a shortest horizontal and vertical path length is returned, *bends*, which returns a path with the least number of bends on the same layer, and *down*, which minimises the distance of corners from the bottom edge of the routing space.

The path shown in Figure 7 a) is found by optimising in the order: *down, distance, bends*. This produces a path which runs as tightly to the bottom edge of the routing space as possible - keeping a layer-spacing from the boundary of the cell. This may be used if the height of a routing cell was at a premium, although the route is far from optimal with respect to Manhattan Distance; this technique can generate contour-following routing patterns similar those produced by Groenveld et al (1987).

The path shown in Figure 7 b) is found by ordering the criteria: *distance, down, bends*. This path is similar to one which can be found by generating a symbolic connection defining its topological relation with the obstacle and calling a vertical compaction process and a partial straightening process. This approach has been advocated by Deutsch (1985). Routing at the mask level has an advantage over a symbolic connection and compaction/straightening process in that the path can be optimised with respect to given criteria without having to use quality measures based on the symbolic level. Figure 7 c) shows the path found

262 Guaranteeing optimality in a gridless router

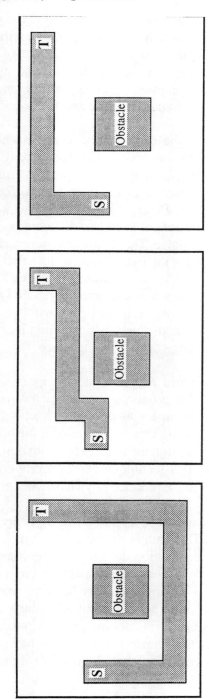

Figure 7 Alternative nesting of optimisation criteria

by ordering the optimisation criteria: *distance, bends, down*. Since *bends* now has priority over *down*, a minimum bend path is found that is higher in the routing space than before. The result in this case is similar to a maximal straightening process, although this again is guaranteed to be optimal with respect to its supplied criteria.

Currently the routing tool also operates to minimise contact count and path resistance and additional criteria can be readily accommodated. These examples indicate that by judicious use of optimisation criteria it is possible to duplicate the track laying strategies advocated by a number of authors. However, using heuristics in this way has the advantage that the final path on silicon is optimal with respect to these criteria as no measurements are taken that are based on potentially misleading information inferred from a symbolic description.

11.5 Operation of the Routing Tool

We first consider the application of the gridless routing tool to the constrained routing problem which was presented in Figure 1 - the example which was used to demonstrate cell stretching caused by the contact positions in a symbolic routing pattern. In this example two interconnection layers are available - metal and polysilicon. Net 2 was declared as a power net requiring a double-width track and the path was optimised against contacts, bends and Manhattan distance. Net 1 was required to be optimised against contacts first, followed by resistance. Nets 3 and 4 were optimised in terms of Manhattan distance first, contacts next and finally the number of bends.

Priority is given to power resulting in net 2 being routed first. Net 1 is routed next as it is the only remaining net with a special rule. The optimisation criteria force a minimum number of contacts and a minimum resistance path. Since both terminals are in metal and there is no possible metal path between them that does not collide with the power track, the path must contain two contacts. In the design rules, polysilicon has a much higher resistivity than metal so the shortest possible polysilicon path should be selected. This leads to a minimal-distance polysilicon underpass under the net 1 track with the remainder of the path being completed in metal.

No rules exist for nets 3 and 4, so the selection of routing order is arbitrary. In this case the nets are routed in a left to right order, net 3 being completed before net 4, leading to the final result shown in Figure 8. This clearly demonstrates the ability of the routing tool to meet the spacing constraints imposed by the functional blocks.

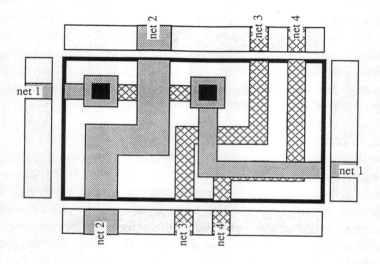

Figure 8 The completed solution to the highly constrained problem

One of the difficulties that exists with conventional channel and switchbox routers is that they are generally inflexible. They are fixed to one style of interconnection, usually with two layers, and are restricted to running horizontal tracks on one layer and vertical tracks on the other. The routing area is usually required to be rectangular and is assumed to be empty.

In order to indicate the level of flexibility that can be obtained by routing using the grid-less single-net router, an example is presented of an L-shape channel with terminals on three layers; two metal and one polysilicon. Additionally, one of the nets is a power net requiring a double-width track in metal2. In order to represent this problem to the router, a boundary rectangle is declared. To represent the L-shape, three rectangular obstacles, which lie vertically above each other on each layer and define the no-go areas, are declared within the problem description file. Terminals are placed on the periphery of the no-go areas and around the boundary rectangle.

In order to complete the routing of the example, the power net was routed first as it has special requirements. The order of the remaining nets was decided by using a heuristic search method. A congestion measure was used to give an estimate of the probability of completing the routing of a partially solved state. This produced the routing result shown in Figure 9 a).

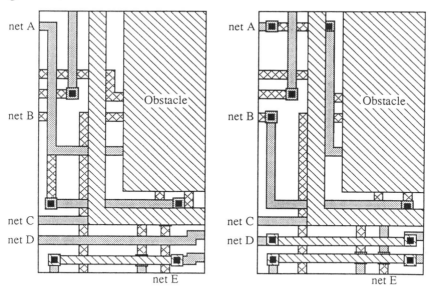

Figure 9 Two solutions to an L-shaped channel problem

One of the advantages of routing at mask-level is that the routing pattern can be changed to fit different fabrication processes. Figure 9 b) shows the same routing example routed using the same ordering of nets. However, in this case, the required separation between features on the metal1 layer has been doubled. When the routing pattern is generated for the new process, the positioning of metal1-metal2 contacts is altered to take account of the more stringent design rules. In particular it can be seen the routing tool modified the contact positions on net B to avoid conflict with net C and similarly on net E to avoid conflict with net B. The tool subsequently found a new path for net A given the changes in net B and the path for net D incorporates layer changes due to the modification in net E. This example clearly demonstrates the flexibility of the routing tool and how the method of mask-level routing can be used to produce layout which best conforms to the constraints of a given fabrication process.

11.6 Conclusions

Looking back to the aims specified in section 11.2, full-custom routing is required to be grid-free and produce a guaranteed, optimum path which conformed to arbitrary process design rules. A flexible routing system has been developed based upon a gridless point-to-point routing tool. This operates in real dimensions and produces mask-level routing patterns from user-specified design rules. The system takes account of special requirements of individual nets and the routing tool is controlled such that multiple optimisation criteria are appropriately balanced for the current task.

Currently the strategies used by the Planner to control the routing tool are relatively simple, based upon assigning priority to special nets and estimating routing congestion. All the routing problems which have been illustrated were soluble and the routing system was successful in finding appropriate interconnections. In practice however the constraints placed by the functional cells on the dimensions of the routing channel may be so severe that no interconnection pattern exists; in this case space has to be created in the routing channel and the abutting cells. Currently the routing channel is simply expanded at the point where failure occurs; further work on the system is targeted towards the development of more sophisticated control strategies and the incorporation of failure analysis.

11.7 Acknowledgement

The authors wish to acknowledge the support to the work described in this chapter made by British Telecom Research Laboratories and the Science and Engineering Research Council.

11.8 References

DEUTSCH D. N. (1985): "Compacted Channel Routing", Proceedings of the IEEE International Conference on Computer-Aided Design, Pp 223-225.

GROENVELD P., CAI H. & DEWILDE P.(1987): "A Contour-based Variable-width Gridless Channel Router", Proceedings of the IEEE International Conference on Computer-Aided Design, Pp 374-377.

HART P. E., NILSSON N J. & RAPHAEL B.(1968): "A Formal Basis for the Heuristic Determination of Minimum-Cost Paths", IEEE Transactions of Systems, Science and Cybernetics, Vol SSc-4, No. 2, Pp 100-107.

JOHANNSEN D.,(1979): "Bristle Blocks: a silicon compiler", Proceeding of the 16th Design Automation Conference, Pp 310-313.

LEE C.Y. (1961): "An Algorithm for path connections and its application", IRE Transactions on Electronic Computers, vol. EC-10, P77p. 346-365.

NACLERIOl N.J., MASUDA S. & NAKAJIMA K.(1987): "Via Minimization for Gridless Layouts", Proceedings of 24th Design Automation Conference, Pp 159-165.

PINTER R.Y (1983): "River Routing" Methodology and Analysis", Proceedings 3rd Caltech Conference on VLSI, Pp 141-163.

REVETT M.C.(1985): "Custom CMOS Design using Hierarchical Floor Planning and Symbolic Cell Layout", Proceedings of the IEEE International Conference on Computer-Aided Design, Pp 146-148.

RIVEST R. L. & FIDDUCIA C. M.(1982): "A 'Greedy' Channel Router", Proceedings of 19th Design Automation Conference, Pp 418-423.

RUBIN F (1974): "The Lee Path-Connection Algorithm", IEEE Transactions on Computers, Vol C-23, No. 9, Pp 907-914.

SATO K., SHIMOYAMA H., NAGAI T., OZAKI M. & YAHARA T.(1980): "A 'Grid-Free' Channel Router", Proceedings of 17th Design Automation Conference, Pp 22-31.

SHARPE M.F. & MACK R.J. (1988): "A Mask-level Plan Generator for use within a Routing Expert System", Proceedings of the 31st Midwest Symposium on Circuits and Systems, Pp 1074-1077.

TAYLOR S.(1984): "Symbolic Layout", VLSI Design, Pp 34-42.

Index

A

A* Algorithm 254
Abstract data types 28
ADFT 111
AI-based design
  techniques 1
AI Paradigm 7
Algebraic techniques 163
Algorithmic thinking 3
ALLOC 36
Allocation 60
Analogue systems 202
Analogue VLSI layout
  synthesis 201
Analysis 10
Area overhead 125, 126, 128, 130
AUTOALL 36
Automatic generation
  of test patterns 163

B

Backtracking 4, 125
Behavioural synthesis 78
BILBOs 132
BIST 125

Boolean difference 163
Boyer Moore theorem
  proover 25
Bus control block 190

C

Case-based plan 226
Case-based planning
  scheme 218
Cell stretching 247
CHIPADE 202
Chip assembly 246
CHIPWISE 209
Clique partitioning 65
Compiler optimisations 57
Component
  orientation 218
Component
  interconnectivity 215
Condition vector node
  encoding 65
Conflicting
  requirements 1
Constraints 3
Contour following 261
Control data flow graph 54
Conventional
  programming

languages 3
Cost/benefit analysis 125
Cost model 129
Cost modelling 126
CRIB Critical paths 106
Crosstalk 214

D

D-algorithm 163
Database 27
Data path assignment 61
DEALLOC 36
DEFN 27
Design and analysis 1
Design assistant 1, 10
Design choices 4
Design costs 130
Design for testability 122, 125, 180
Design iterations 2, 4
Digital systems 1
Direct access 190

E

EALS 203
Early design choices 1
Economics model 129
Error Decoder 37
ESPRIT 140
Events 27
EXIT 36
Expert analogue system layout 201, 240
Expert hardware design system 10
Expert system 1, 96, 125
Explanation 1

Explanation of the design decisions 17
Explanatory/interface module 97
Extracting tests from FAFs 174

F

Fast discrete cosine transform 76
Fault augmented function (FAF) 164
Fault cover 128
First order logic 27
FIS 107
Floorplanning 62, 245
Formal specifications 25
Frame-based 97, 98
Frontier list 254
Full custom 245
Full scan 132
Functional blocks 124
Functional cells 202
Functional module 205, 208

G

Generated test plan 194
Graph colouring 65
Grouping of components 223
Guiding test plan 193

H

Hardware description language 26
Heuristic knowledge 108
Heuristics 7, 9

Hierarchical clustering 64
Hierarchical designs 192
Hierarchical test 142
Hierarchy 181
High level design
  decisions 13
High level synthesis 26, 52
HITEST 103
HOL 25
Hypothesis 28

I

IKBS 1, 9
ILAC 202
Ill-structured problem 1
Incomplete initial
  specifications 7
Incomplete requirements 4
Inconsistencies 7
Inference engine 97, 98
Initial test plan 193
Intelligent CAD systems 4
Interacting objectives 6
Interactive design tool 24
Interactive interface 34
Interconnections and
  constraints 203
Interface synthesis 51
I-paths 125
Isolation 213

K

KBES 9
Knowledge acquisition
  module 97
Knowledge-base 97, 124,
  128, 129, 137

Knowledge-based
  expert system 9
Knowledge-based
  method 9
Knowledge-based
  systems 124
Knowledge database 206

L

Layer type 234, 235
Layer type information 209
Leaf-macros 183
Libraries 33
Library of existing
  designs 13
Logic-based 97

M

Manage the complexibility 1
Marketing 129
Manhattan paths 257
Matching 213
Microprocessor based
  systems 12
Min-cut partitioning 64
MIND 106
Mixed integer linear
  programming 65
Modified design 5
Module assignment 61
Modules 202
MRG/SPL 36
Multiple connection path
  case-based planner 231
Multiple strategy systems 124

N

Netlist information 206

O

Object-oriented programming (OOP) 18
Op Amp 234
OPEC2 113
Operator assignment 61
Optimisation 139

P

Packaging costs 130
Partial solution 7
Parasitic 214
Parasitics 227
Parsing operation 206
Piramid VLSI design system 186
PLA 125, 132, 137
Placement 203, 208, 209
PLA-ESS 112
Pre-determined strategy 3
Problem space 7
Procedural knowledge 108
Production costs 130
PROLOG 206
PROVE-LEMMA 27
Pruning the search space 9

Q

Quantifier free 27
QUOTIENT 29
QUOTIENT-TIMES-TIMES 29

R

RAM 132
Reasoning strategies 98
Reduction rules 167
Register assignment 61
Repetitive structure 36
Requirements capture 11
Re-use 13
REWRITE 36
Re-write rule 28
Routing 203, 208
Routing overhead 128
Ruled-based 94
Rule-based expert system 213
Rules of thumb 9

S

SAGE 26
SALIM 202
Saticficing 6
Scan 124
Scan chain 190
SCIRTSS 105
Scheduling 59
Scheduling and allocation 26
Self-test 132
Shell principle 28
Simulated annealing 76
Solution proposals 8
Sphinx system 186
Standard cell 126

STICKS 209, 240
Stick representation 245
STORE 35
STRICT 25
Structural knowledge 108
System level synthesis 51
Synthesis 10, 125
SUPERCAT 102

T

TDES 109. 124
Testable design
  methodologies 124
Testability 122, 125, 180
Testability analysis 123
Testability synthesis 193
Testability rules 187
Test application
  protocol 186
Test control block 191
Test costs 132
Test inference scan-cell 190
Testing costs 122
Test generation 101
Test generator 12
Test pattern assembler 196
Test pattern generation 128
Test plan 185
Test strategy 129
Test strategy controller 128, 137
Test strategy planning 122, 124
Test synthesis 193
Test time 125, 126
TIGER 110, 124
Timing analysis 14

Top-level requirements 13
Trade-offs 11
Transfer paths 190
Transformation 59
Transformational
  correctness 123
Transformational
  synthesis 122
Transparency 139

U

Unbounded search
  space 8
Understandability 1

V

VHDL 83
VHDL for synthesis 67
VLSI design 24

W

Weighting value 209, 214
Weights 126

Z

Zissos 167